WORLD CITY

WORLD CITY

DOREEN MASSEY

polity

First published in 2007 by Polity Press

Reprinted 2007 (twice), 2008 (twice), 2010, 2011, 2012, 2013, 2014, 2015, 2016

Polity Press
65 Bridge Street
Cambridge CB2 1UR, UK

Polity Press
350 Main Street
Malden, MA 02148, USA

ISBN-13: 978-07456-4059-4
ISBN-13: 978-07456-4060-0 (pb)

A catalogue record for this book is available from the British Library.

Typeset in 10.75 on 14 pt in Adobe Sabon
by Servis Filmsetting Ltd, Manchester
Printed and bound in the United States by RR Donnelley

The publisher has used its best endeavours to ensure that the URLs for
external websites referred to in this book are correct and active at the time
of going to press. However, the publisher has no responsibility for the
websites and can make no guarantee that a site will remain live or that the
content is or will remain appropriate.

Every effort has been made to trace all copyright holders, but if any have
been inadvertently overlooked the publisher will be pleased to include any
necessary credits in any subsequent reprint or edition.

For further information on Polity, visit our website: www.polity.books.com

CONTENTS

ACKNOWLEDGEMENTS

The arguments here have benefited from conversations, debates and political engagements with many people over many years. My efforts to pull them into some kind of shape for this book have also been greatly aided by others. Ellie Jupp provided early research assistance, organising my heaps of cuttings, hunting down documents, getting hold of statistics, and providing comments. Her efficiency set me on my way. A number of people read all or part of the manuscript, at various stages, and/or provided invaluable comments and clarifications – thank you to Allan Cochrane, Jane Wills, Ray Hudson, Adam Tickell, Ken Livingstone, Mick Dunford and Ian Gordon. Other friends working in and on London, academically and politically, have been a source of ideas and information both over the years and in the immediate preparation of this book. Way back in the 1980s there was the Ariel Road Group (Maureen Mackintosh, Hilary Wainwright, Michael Ward, Vella Pillay, Ken Livingstone, Robin Murray and Michael Rustin) that met in my flat once a month after the demise of the GLC to try to keep the ideas flowing and to begin to take them further into new times. Later, long discussions with Ash Amin and Nigel Thrift (which resulted in joint writing of our own) began the development of some of the ideas in Part II. I have learned much too from friends in more recent groupings, especially the London and European Social Forums.

Neeru Thakrar helped me with the initial stages of the physical preparation of the book. Throughout the latter part of writing, the help of Angela Daniels has been utterly invaluable – producing the typescript, chasing up queries, and being a constant source of support. It has been great working together. Finally, Caroline Richmond helped me through copy-editing with precision and warmth, and Emma Hutchinson at Polity, too, beautifully combined friendliness and efficiency. My thanks to all.

I am grateful to Jonathan Freedland and *The Guardian* for permission to reprint a substantial part of his article 'It may be beyond passé – but we'll have to do something about the rich' (23 November 2005; copyright Guardian News & Media Ltd 2005); to *The Guardian* for permission to quote from their leader 'Unbalanced Britain' of 6 February 2003 (copyright Guardian News & Media Ltd 2003) and Larry Elliott's 'The United Kingdom of London' (5 July 2004; copyright Guardian News & Media Ltd 2004); to *The Observer* for permission to quote from Nick Cohen's 'Without prejudice' (22 February 2004; copyright Guardian News & Media Ltd 2004) and to the *Evening Standard* for permission to quote from C. Freeman's 'Mayor's funding plea "will divide Britain"' (26 June 2001). Jane Wills granted permission for me to use the information on the London Living Wage campaign shown in table 6.1, the Carbon Web illustration (figure 10.1) is courtesy of PLATFORM, and Ken Livingstone granted permission to quote from the press releases of 7 and 8 July 2005.

PREFACE

AFTER THE CRASH

This book was first published in July 2007. Its aim was to point to the dangers of the dominance of the finance sector and its surrounding constellation of activities, in London, in the UK, and internationally. It was on the basis of finance that London reinvented itself from the 1980s on. It was around this sector that emerged a remoulded social stratum of the super-rich. It was on the basis of the financial sector that the British economy was levered out of its long decline. And it was around this economic basis, too, that a new social settlement, commonly called 'neoliberal', was built and became hegemonic, its tenets embedding themselves deeply into popular common sense. Wider still, it was on the basis of the financial sector that London asserted a new imperial role. It was all so self-assured. It was a brave new world. This is the world that is scrutinised in this book.

That world has now imploded, and one of the central implications of the arguments here is that we must not go back to business as usual.

Cities (and especially those called 'world cities') were central to this social settlement; in their glitz, their gentrification, and their acute inequality, they are home base to a new global elite – London perhaps above all. As this book argues, London, or more precisely its financial constellation, was at the heart of the establishment of neoliberalism as hegemonic. Its fortunes were built on deregulation, privatisation and marketisation, and it was these forces that spread through the country and around the world.

'After the crash' this argument can be extended. For if London was at the centre of this kind of society in its pomp, then it is also at the crux of its crisis. It is often said that this crisis is global, or that it was made in the USA. Well, it was certainly triggered in the USA, through the collapse of the sub-prime markets, but an immediate trigger is not the same as underlying causal conditions. Likewise, the crisis is certainly global, in that it had effects around the world. But, as this book argues, globalisation is made in places. The global is grounded. And one of the key localities where financial globalisation was invented and orchestrated was London.

One question that arises now is: Is this 'just' a banking crisis or could it precipitate a wider shift in social forces? Could it herald the beginning of the crumbling of the social settlement that has characterised the last thirty years?

The 1980s, the decade when this social settlement was finally established, was a period of social contest – over Thatcherite cuts, over the closure of the mines and the decimation of manufacturing, over the government of London and other rebellious local authorities. That is how shifts between social settlements happen. They are moments when the future seems open.

Are we now – potentially – at another such moment? Is the dominance of finance now less assured? Could its crisis lead, as on occasions it has promised to do, to a wider questioning: of the rule of untramelled market forces, of growing inequality, and of a dominant culture of greed and self-interest? These are questions that will not be settled quickly.

The economic basis for the neoliberal social settlement was laid down under the Thatcher government, through the accelerated decline of mining and manufacturing and the rise to dominance of the City. New Labour accepted this inheritance, and used the taxes flowing from finance into the exchequer to fund its social-democratic programme through a public sector itself culturally remoulded to reflect the ideological tenets of the market. Thus was the economic settlement embedded in society more widely as an unquestioned common sense.

Even at its height, this settlement was fraught with problems. Finance proclaimed itself the golden goose, but its preference was to invest in assets rather than productive activities. Its very dominance made life more difficult for other parts of the economy. Inequality soared and could not be reined back even by a plethora of anti-poverty programmes, as the very rich grew in numbers and in wealth. The acute inequalities in London made running the city deeply problematical. The gap between North and South of the country widened as growth was concentrated in London/the South-East, and resources and skills followed. It was a pattern propagated around the world, as inequalities sharpened both between and within countries. All this is analysed in the pages that follow. But those pages were written as a trenchant critique of an economy and society widely judged, in spite of everything, to be 'successful'. Today they must be read as a dire warning that, whatever way is found out of the financial crisis, we must not simply reconstruct those times.

Indeed, in many ways we cannot. That model is now broken. The dominant cry now is that 'the public deficit' must be reduced – that public expenditure must be cut back. It is a cry that has been forcefully contested. But it is what 'the markets' demand – the same markets whose behaviour produced the implosion and whose failure necessitated their rescue by the public. The result could be yet further inequality, nationally, within London, and between North and South. There are gender implications too, for, while it is women who tend to suffer more from cuts in public expenditure (through loss of jobs as well as of services), what is clear from the analysis in this book is that the group statistically most responsible for the sharp inequality resulting from thirty years of neoliberal growth is highly paid men.

'The markets' that demand such cuts are not some impersonal force of nature. They are the same social strata that, over the last three decades, have refused to be a serious motor of the economy and have been at the centre of the creed of selfish individualism. And now they take their bonuses again, and presume to tell us what to do. There are serious questions here of democracy itself.

There is, nonetheless, and as a result of the crisis, a widespread recognition that the finance sector, especially in its wilder reaches, must be made more accountable and its regulation strengthened. One clear message of this book, however, is that while this is necessary it will not be enough. What also needs to be addressed is the sheer size and dominance of this sector, both in the national economy and in the economy of London. And that will mean challenging its hegemonic stories. As this book shows, much work is put in to ensure the popular currency of these stories. Equivalent work is needed to address them: that finance is the golden goose without which the economy would crumble, that it is the motor of a productive economy, that if any attempt is made to intervene it will leave (it doesn't actually go, and it is a moot point whether anyway that would in the long term be such a loss). These stories are already challenged in the pages that follow.

And what of London in all of this? The national economic model was accepted as inevitable in London too – that the City would be the leading sector – and, while this book was written when Ken Livingstone was mayor, the shift to Boris Johnson did not alter that particular prioritisation. Yet, as it is argued here, London is in many ways a progressive city, and the dominant presence within it of one of the founding localities of global neoliberalism has been an anomaly from which most political debate on the left has averted its gaze. Now, with the (potential) puncturing of the hubris of these social strata, will it be more possible to challenge this shape of the city? Can we question the identity of place itself?

London is also a world city in that it has effects on the wider planet beyond it, and it is argued here that we can build a responsible 'politics of place beyond place' that asks serious questions about the global impact of the local – a global ethics of the local place. Here, too, both the importance and the urgency of the argument have been dramatically reinforced by the way in which the financial crisis rocketed around the world, wreaking havoc in so many places so far away.

Both because it is a hearth of neoliberalism and because it is basically a radical and progressive place, the contest over London *matters*. In the 1980s, when the neoliberal hegemony and the dominance of finance were still being contested, London was at the heart of the political confrontation, not least in the battle between the Thatcher government and the Greater London Council, with London's voice demonstrating a way out to the left. What will London's voice be raised for now? One of the questions this book persistently asks (and asks us all to ask) is: What does this place stand for?

But while the text starts in London, and while it is from this city that many of its key themes emerge, it is not about London alone. These are questions that face us all.

Running through this book is the term 'neoliberalism'. It is the name we have frequently given to the social settlement of the last thirty years. So, when the banks failed and had to be bailed out by the state, it was easy to point to the logical contradiction. Some even pointed, perhaps in hope, to the *end* of neoliberalism. But things are not so simple. First, neoliberal thinking achieved an astonishing hegemony, and its tenets – the naturalness of market forces, the inevitability of individual self-interest, the negative attitude to state intervention – still run deep in popular consciousness. *Political* arguments are about more than economic logic. Second, as this book shows, the nostrums of neoliberalism were always applied selectively, when useful, and disregarded when not. Neoliberalism as an economic doctrine has been a legitimating tool in the armoury in what is, at bottom, a submerged contest between social forces. It is with this contest, within London, in the country as a whole, and in the global ramification of these, that this book is fundamentally concerned.

So, turning the page now means plunging back into that world when finance was in its imperious pomp. London had just won the competition to hold the Olympics. But it had also just been bombed.

London, June 2010

INTRODUCTION: 'THE FUTURE OF OUR WORLD'?

In the numbed days after the first bombs went off on London's public transport, in July 2005, Ken Livingstone, the city's mayor, said: 'this city is the future' (GLA, 2005b). 'This city', he said, 'typifies what I believe is the future of the human race and a future where we grow together and we share and we learn from each other' (ibid.). He set London in the context of cities around the world, and of a longer history:

> If you go back a couple of hundred years to when the European cities really started to grow and peasants left the land to seek their future in the cities there was a saying that 'city air makes you free'[,] and the people who have come to London [–] all races, creeds and colours [–] have come for that. This is a city [where] you can be yourself as long as you don't harm anyone else. You can live your life as you choose to do rather than as somebody else tells you to do. It is a city in which you can achieve your potential. It is our strength and that is what the bombers seek to destroy. They fear

freedom, they fear a world in which the individual makes their own life choices and their own moral value judgements and that is what they seek to snuff out. But they will fail.

This year for the first time in human history a majority of people live in cities. London continues to grow and I say to those who planned this dreadful attack[,] whether they are still here in hiding or somewhere abroad, watch next week as we bury our dead and mourn them, but see also in those same days new people coming to this city to make it their home to call themselves Londoners and doing it because of that freedom to be themselves. (Ibid.)

On the day itself, speaking from Singapore where London had just won the competition to hold the Olympics in 2012, he had addressed not only Londoners but the world at large:

I want to say one thing specifically to the world today. This was not a terrorist attack against the mighty and powerful. It was not aimed at Presidents or Prime Ministers.[1] It was aimed at ordinary, working-class Londoners, black and white, Muslim and Christian, Hindu and Jew, young and old.

. . . we know what the objective is. They seek to divide Londoners. They seek to turn Londoners against each other. I said yesterday to the International Olympic Committee, that the city of London is the greatest in the world, because everybody lives side by side in harmony. Londoners will not be divided by this cowardly attack. They will stand together in solidarity alongside those who have been injured and those who have been bereaved[,] and that is why I'm proud to be the mayor of that city. (GLA, 2005a)

And he had addressed 'those who came to London today to take life':

I know that you personally do not fear giving up your own life in order to take others – that is why you are so dangerous. But I know you fear that you may fail in your long-term objective to destroy our free society and I can show you why you will fail.

In the days that follow look at our airports, look at our sea ports and look at our railway stations and, even after your cowardly attack, you will see that people from the rest of Britain, people from around the world[,] will arrive in London to become Londoners and to fulfil their dreams and achieve their potential.

They choose to come to London, as so many have come before[,] because they come to be free, they come to live the life they choose, they come to be able to be themselves. They flee you because you tell them how they should live. They don't want that[,] and nothing you do, however many of us you kill, will stop that flight to our city where freedom is strong and where people can live in harmony with one another. Whatever you do, however many you kill, you will fail. (Ibid.)

The Olympic bid had emphasised, above all, London's ethnic and cultural diversity. A host of messages, on websites and posters, on banners and in interviews with people on the street, said the same thing. At a gathering of 25,000 people in Trafalgar Square a week later Ben Okri read a poem he had retitled 'A hymn to London': 'Here lives the great music of humanity' (*Evening Standard*, 15 July 2005, p. 6). And Ken's entry in the Book of Condolence read: 'The city will endure. It is the future of our world. Tolerance and change' (*The Guardian*, 12 July 2005, p. 3).

With the *Evening Standard* (the city's newspaper) running a special issue on 21 July with the banner headline 'London United' and distributing a poster 'London stands united', and

Time Out ('London's weekly listings bible') running a front cover that said simply 'Our City' (13–20 July edition), it was plain that what was being proclaimed here was an identity of place; a London 'we'. Moreover, while there is little doubt that the particular events reinforced a sense of identity, and of this particular identity, it was not a brand-new construction. In the spring of the previous year, MORI had published the results of an enquiry into 'What is a Londoner?'. As well as demonstrating high degrees of both identification as Londoners and satisfaction with London as a place to live (with, in both cases, ethnic minorities being the most positive), the interviews also revealed cultural diversity and cosmopolitanism together as the things that most made people 'proud of London' (MORI, 2004).[2] In January of 2005 a special supplement of *The Guardian* newspaper had been published: *London: the world in one city: A special celebration of the most cosmopolitan place on earth* (*The Guardian*, 2005). This last, perhaps inevitably, tended to map different groups, so-called communities, and this is indeed one aspect of what is at issue. More importantly, what was being hailed in July 2005 was a 'mixity', so to speak, of lived practices, the criss-crossing multiple allegiances described by Saghal and Yuval-Davis (2006) and what Gilroy (2004) has described as a convivial demotic cosmopolitanism, rather than, as 'multiculturalism' is sometimes understood, the juxtaposition of, and negotiated relations between, mutually boxed-in communities.

This, then, was a claim to place identity precisely as a constellation that might problematise inside and outside: 'London is the whole world in one city' (Livingstone, in *Time Out*, 13–20 July 2005, p. 3). It was a claim to 'place' as open rather than bounded, as hospitable rather than exclusive and excluding; to place as ever changing rather than eternal. Place as a constellation of trajectories; as a *meeting*-place (Massey, 2005). The city as *ville-franche* (Derrida,

2001). And it was a claim to place identity made not only by the powerful but by many on the streets as well.

It was in this guise that London was being celebrated as a world city. All of these claims – to specificity, to unity, to holding out a future for the world – were built round the rich ethnic and cultural diversity of London. Livingstone's passion, in particular, sounded out in stark contrast to the manufactured sincerity of Tony Blair. Nor did Ken speak of good and evil, but of real grounded politics. His commitment to diversity and hospitality rang a clear note after a general election, some months previously, in which dismally negative debates about immigration and asylum had been prominent.

The claiming of this place identity also threw up bigger questions, not least that of its validity (Keith and Cross, 1993).[3] Although there were references to 'harmony' and simple 'unity', this wasn't, at least at an explicitly political level, about some unproblematical, happy-clappy diversity. Indeed, in the days that followed, 'multiculturalism' was a contested term. At the end of July, there was a thoughtful debate between Jonathan Freedland, a *Guardian* columnist, and Livingstone in which Freedland criticised Livingstone (whom he had praised for his initial 'healing' words) for introducing divides into this diversity. The accusation was of being too accommodating to 'contemporary Islam', and in particular Sheikh Yusuf al-Qaradawi, and of taking sides on the Palestine/Israel question. '"They seek to turn Londoners against each other," Livingstone said of the terrorists on July 7. Yet what was he doing last week?' (Freedland, 2005b). The following week, in an article of his own, Livingstone wrote of how London could make itself safer. Among his proposals (which included withdrawal from Iraq) was the need to 'shrink the pool of the alienated that bombers draw on by treating all communities as equal parts of British society – not only theoretically, but in reality' (Livingstone, 2005). He makes clear that he disagrees with Qaradawi on a whole range of issues but that the latter should not – any more than Ariel Sharon – be banned from entering the country: 'It is impossible to say that Britain's Muslims should be treated with respect but that their religion's most eminent representatives must be banned.' The key word is respect.

This, then, is an attempt to construct not a bland diversity so much as a recognition of differences with all their conflicts and problematical implications. It does not mean not being critical, or not taking up clear political positions. It recognises that this may be a conflictual negotiation of place. But it insists that participants should both be allowed in and treated with respect. It is a thoughtful political position.

However, that in itself means that this is a *particular* position. It is secular; it is Western. Indeed, Livingstone specifically claims it as such, naming it 'the best of the West' (*Evening Standard*, 21 July 2005, p. 13). It assumes a framework of negotiation and respect. If multiculturalism is a universalism (as is sometimes suggested) then there are other universalisms which might differ from it, even oppose it, even be a source of the bombings themselves. 'London represents the best of the West, and for that reason alone it is a target for terrorism' (ibid.). The opposition to liberalism from Sayyid Qutb to bin Laden is well documented (for one analysis see Boal et al., 2005). Moreover, even multiculturalisms develop in particular guises and within particular hegemonic assumptions that frame its working (Hall, 2000; Hesse, 2000; Mitchell, 2004; Nash, 2005). Indeed, only given some hegemonic assumptions (always themselves negotiated and open to further negotiation) can the negotiation of place take place.[4]

Claiming London as the future, then, is actually on this logic to claim it as *one* potential future among futures. It is not just a description, or a claim to be at the front of the historical queue. It means, rather, that London *stands* for something, a particular kind of future, but it also carries the possibility that this may be one future in a still varied and plural world. Maybe other places, other cities, will be different.

There is a long tradition of trying to dragoon cities into a singular linear history (from Athens to Los Angeles, or some such). It was never a story that was capable of capturing, of recognising, the multiplicity of the world. (What of Samarkand? What of Tenochtitlán?) The framing of London as 'standing for something' also, actually, leaves open the possibility that the future will be equally multipolar. London stands for one kind of future, but there may be others.

* * *

All of this reflection and establishment of place identity was framed within a geographical imagination of London as a world city. But ethnic and cultural mixity is only one aspect of its being a world city, only one aspect of the city's relationship with the world. For one thing, it focuses on the internal, on people arriving (the world coming to *it*). And this is indeed one side of 'a global sense of place' (Massey, 1991). But world cities, as indeed all places, also have lines that run out from them: trade routes, investments, political and cultural influences, the outward connections of the internal multiplicity itself; power relations of all sorts that run around the globe and that link the fate of other places to what is done in London. This is the other geography, the external geography if you like, of a global sense of place. For each place this geography, this tentacular stretching of power relations, will be particular. For London, precisely as a world city, this is significant, not just for the metropolis itself (it could not survive for a day without the rest of the world) but also because of the effects it has even in the remotest corner of the globe.

The response to the bombings did not touch on this other aspect of being a world city. Yet as well as being so ethnically mixed, London is also a seat of power – political, institutional, economic, cultural. Its influences and its effects spread nationally and globally. It is a heartland of that socio-political economic formation that goes by the name of neoliberalism. It was forces in London, articulated above all around the financial City (capital C), that took the lead in advocating and developing the deregulation that lies at the heart of that formation; and it is a command centre, place of orchestration, and significant beneficiary of its continuing operation. This city stands, then, as a crucial node in the production of what is an increasingly unequal world. Economic inequality has over recent decades and on most measures increased globally. It has also increased nationally; within the UK the old 'North–South divide' has widened (Adams, Robinson and Vigor, 2003), and has increasingly taken the form of an ever-expanding London 'versus' (we shall come back to that) what is usually, tellingly, called 'the rest' of the country. And London itself, already the most unequal region in the nation, is becoming increasingly so, in terms both of earnings and of income (Hamnett, 2003). Livingstone was surely right, in defiant response to the bombings, that people flock to London and will continue to do so because of the freedom it offers them 'to be themselves'. But people find their way here for other reasons too. They come because of poverty and because their livelihoods have disappeared in the maelstrom of neoliberal globalisation. They come out of destitution and desperation (and millions more are left behind). And it has to be at least a question as to whether London is a seat of some of the causes of these things.

Moreover it is not only into London that people are crowding. Livingstone cites the now well-known fact that 'for the first time in human history a majority of people live in cities'. However, the biggest cities, and those growing the fastest, are in the global South. And in these cities, in Asia and in sub-Saharan Africa, over half the people live in slums (UN-Habitat, 2003; 2004) in conditions that Davis (2004, 2006) has documented with such power. Ken Livingstone spoke, thoughtfully and correctly, of the need to shrink the pool of the alienated in London and in the UK as a whole. Others, writing in the aftermath of the attacks on New York, have reflected on the festering anger in cities of the majority world, what they call 'the slum conurbations of the World Bank world': 'never before – this is the truly chilling reality – have the wretched of the earth existed in such a bewildering and enraging hybrid state, with the imagery of consumer contentment piped direct into slum dormitories rented out by the night, at cutthroat prices, to hopelessly indebted neo-serfs' (Boal et al., 2005, p. 173).

Cities are central to neoliberal globalisation. The increasing concentration of humanity within them is in part a product of it. Their internal forms reflect its market dynamics (the shining spectacular projects, the juxtaposition of greed and need). The competition between them is both product and support of the neoliberal agenda. And in certain cities (those we call world cities, or global cities) is concentrated the institutional and cultural infrastructure that is key to all of this. It is not just 'that neoliberalism affects cities, but also that cities have become key institutional arenas in and through which neoliberalism is itself evolving' (Brenner and Theodore, 2002, p. 345; Sassen, 1991). Cities, then, are crucial to neoliberal globalisation but they figure in very diverse ways within it. London as a centre of command and orchestration and as, indeed, a focus of migration and a home to an astonishing multiplicity of ethnicities and cultures is a part, and a powerful part, of the same dynamics that produce, elsewhere in other cities, Davis's 'planet of slums' where, 'instead of becoming a focus for growth and prosperity, the cities have become a dumping ground for a surplus population working in unskilled, unprotected and low wage informal service industries and trades' (UN-Habitat, 2003, p. 46). Is this then another side of London as 'the future of the world'? Does London also stand for this?

*   *   *

'What does this place stand for?' is a question that can and should be asked of any place. Its import and urgency will vary between places (global cities may have more possibility in the sense of room for manoeuvre, and more responsibility in the sense of the magnitude of their effects), but it is a question that makes each and every place a potential arena for political contest about its answer. The constraints are undeniable (from the global movements of capital to the corsets imposed by national policy), but there are possibilities for responses that question and even rework and undermine those constraints. Conceptually, it is important to recognise that the global is as much locally produced as vice versa, that an imaginary of big binaries of us and them (often aligned with local and global) is both politically disabling and exonerating of our own (and our own local place's) implication, and that the very fact of specificity (that places vary) both opens up the space for debate and enjoins us to invent. Moreover, it will be argued, not only is it politically possible, it is also a political responsibility, to find some way of addressing that question. It is a challenge not only for the local state, but for the grass roots of the city too, indeed for all those who in one way or another take a part of their identity from the fact that they are here.

*   *   *

In this more complex picture, then, London's character as an image of a future world is at least ambiguous. Internally, too, like most cities, it is both enormously pleasurable and a site of serious deprivation and despair. The account presented here attempts to weave a course between on the one hand the dystopian visions and apocalyptic urban accounts, generalised perhaps from experience in the USA (and further generalised through the power-geometries of academe and the publishing industry), and on the other hand those overeasy skateboarding celebrations of cities as pure fun. In part, this is a general position (few cities if any are solely one or the other); moreover an insistence on complexity leaves open more opportunities for politics. In part too, and for similar reasons, although all the arguments here are generalisable, I want also to insist on specificity. London is, indeed, even more ambiguous than already described. This is a metropolis that, drawing on its history as imperial capital, is now a seat of neoliberal globalisation, but also one that has twice elected a mayor who lambasts global capitalism. (When newspapers seized upon this during his second electoral campaign, presumably aiming to indict him with 'loony-leftism', his popularity increased still further.) 'Neoliberalism' is sometimes written about as though there is an automatic transmission belt from some ethereal sphere of greater forces to 'how it plays out on the ground'. It is not so. There are indeed pressures and constraints, often of immense power, but there are also agents who play along, or resist, or struggle mightily. There is room for political intervention. The urban 'is not a policy area in which outcomes are given, in which a single agenda is being or can be forced through. It relies on continuing the construction of different visions for the city, which also turn out to be different visions for the wider society' (Cochrane, 2007, p. 145). The current administration of London is, in many ways, radical and left-leaning. The first years were expended upon a highly political, popular, though ultimately unsuccessful campaign to prevent the New Labour national government from privatising elements of the provision of underground transport. There are campaigns against racism, redistributive policies on transport and on affordable housing, and radical approaches to climate change and a wide range of environmental issues. There is a genuine concern about the poverty and inequality within the city. When George W. Bush came to London (in its guise as seat of the national government) the mayor quite explicitly did not welcome him. On the contrary, in 2006 he welcomed, as a result of a specific and personal invitation, Hugo Chávez of Venezuela. And of course there has been the stance against the invasion of Iraq – 2 million people on the streets and the mayor on the platform. And the city is full of grass-roots campaigns. This is no simple transmission belt for neoliberalism. And yet it is made here.

* * *

This book is centred on London. It is not concerned with the detailed documentation of the city. There are some wonderful books that do this (Buck et al., 2002; Hamnett, 2003; for instance) and I draw on them in the argument here. The aim of this book is to stand back a bit and explore some wider questions. If this city is in any sense a herald of the future, these are questions that must be addressed. And they are questions that affect far more than London. They do this *both* because, as a global city, what happens in London affects far more than London *and* because they are questions that should be posed to all those places – and they are many, for going global has become a universal urban imperative – that aim or claim to be world cities too.

So, if this book is centred on London it is not really only *about* London. It is an essay, rather, that arises *from* London. For one thing, there is the question, anyway, 'what *is*

London?' In its normal guise this is an extremely tedious question. There is a ritual, well established, which occupies every conference about this city, or about the geography of the UK, for at least half a session. 'But where does London end?', someone will cry, to nods and murmers of assent that this is indeed a serious issue. Certainly London spreads beyond its administrative boundaries; probably it reaches to Bristol and Cambridge; maybe from the Severn to the Humber . . . The question in such conferences is always and only: where do we draw the line around this city? And it may be important to ask such a question for a whole host of purposes.

But maybe space, or geography, does not work like that any more (if it ever did). Maybe places do not lend themselves to having lines drawn around them. (London is an extreme example – a good laboratory for the argument – but this is a general point.) A high proportion of Londoners were born outside the administrative boundaries. But it is far more than this. There is a vast geography of dependencies, relations and effects that spreads out from here around the globe. This is not to slide into some easy declaration that 'everyone is a Londoner', but it is to argue that, in considering the politics and the practices, and the very character, of this place, it is necessary to follow also the lines of its engagement with elsewhere. Such lines of engagement are both part of what makes it what it is, and part of its effects. Questions of identity run throughout this book.

Most books on places stay within the place. Peter Ackroyd's huge tome (2000), a 'biography' of London, evokes the voraciousness of the metropolis, its lifespring in profit and speculation, its spewing out of huge amounts of waste, the dominance of many London trades even in the nineteenth century by immigrants from the rest of the country. It focuses on the past of the city and on its internal *genius loci*.

But it does not enquire about the effects of all this elsewhere. Iain Sinclair, in *Lights out for the territory* (1997) and a host of other writings, draws out another, stranger, past. But what of those wider, and stranger, geographies of the present day? The response to the bombings celebrated the internal diversity, emphasised the peoples that come to London; it focused on (these elements of) the city's internal structure. But, as already argued, there is another, wider, geography to being a world city.

There is a disjunction here. On the one hand space and places are increasingly the product of global flows; on the other hand we work with a politics both official and unofficial that is framed by a territorial imagination and formal structure. It is a disjunction that is disabling (to some) and highly useful (to others), and the distribution of that spatial entrapment and enablement varies, from situation to situation. It can be associated with closure, competition, and the evocation of external enemies. Trades-union and other working-class-based struggles can find themselves caught, and thereby weakened, in the us–them structure of, for instance, North *versus* South. The organisations of the financial City and other elements of capital, on the other hand, construct an identity for London, a London 'we', that serves their purposes, not only setting 'London's' interests against others' but also thereby covering over the yawning inequalities within London itself (and indeed on occasions *using* that very inequality to argue that 'London' needs more resources from the national treasury). Meanwhile a left-leaning administration needs those resources but does not want either to alienate the most powerful local forces or to set itself against the working class in other regions. This is just one example, explored in Part II, of another theme that runs through this book – the jarring of a territorialised politics with another geography of flows and interconnections.

It is a dislocation that points above all to a need to build a 'local' politics that thinks beyond the local. What is developed here is an argument *against* localism but *for* a politics of place. There is a need to rethink the 'place' of the local and to explore how we can rearticulate a politics of place that both meets the challenges of a space of flows and addresses head-on the responsibilities of 'powerful places' such as global cities. Such cities raise questions about 'local politics' that are quite different from the more usual 'local *versus* global' framings of the politics of place. They are issues that have been raised for me, again and again, by working in and reflecting on London politics. And the geography of the politics they raise is set, not only in the context of London within the United Kingdom (Part II), but also in that wider spatiality of London in the global world (Part III). What is needed is a politics *of* place *beyond* place.

Actions in one place affect other places. Places are not only the recipients of the effects of global forces, they are – in cases such as London most certainly – the origin and propagator of them too, and this raises the question of responsibility, and specifically responsibility beyond place. If actions and policies adopted within one place negatively affect people elsewhere, what responsibility is involved, and what accountability? If a place's very character is integral to sets of relations at the other end of which is produced poverty or deprivation, how should this be addressed? If the economic sectors upon which the local economy of a place is founded entail unequal relations with elsewhere, with other places, how can this be acknowledged? If the reproduction of life in a place, from its most spectacular manifestations to its daily mundanities, is dependent upon poverty, say, or the denial of political rights, elsewhere, then should (or *how* should) a 'local' politics confront this?

Such questions can be asked of any place. But they are peculiarly urgent in global cities. The idea that the local is a product of the global has become common currency (and this is indeed one aspect of what must be addressed) but it is less often recognised that the global is also, conversely, locally produced. 'The global' so often is imagined, implicitly, as somehow always out there, or even up there, but as always somewhere else in its origins. In fact it exists in very concrete forms in local places. And some places more than others are home-bases for the organisation of the current form of globalisation. London is such a place.

Behind those issues of politics lies a changing world and national geography. In the world as a whole big cities are increasingly dominant. In the United Kingdom, London increasingly overshadows everywhere else. These are huge changes, but they are barely addressed by conventional politics. Within the UK, as will be argued, national government policy has been to acquiesce in and also in numerous ways to feed this voracious growth. Is this the geography we want? The question is rarely posed explicitly in democratic debate. On the planet as a whole the World Bank, one of the institutions that has pursued the policies that contributed to this massive flow of people into cities, has argued that it is through competitive cities that nations as a whole can develop (World Bank, 2000). And cities, of course, compete with each other; it is now virtually *de rigueur* to claim, or to have as an aim, to be in some way or other a 'world city'. It is not clear that this is what people would desire, were they to be asked. Not only are the problems of world cities increasingly under scrutiny (see especially the work of Saskia Sassen), but when cities are ranked in terms of liveability it is not the major global cities that come out on top. At the very least, the question needs to be posed.

These cities have mushroomed in a specific context – that of the shift, from the 1980s onwards, towards neoliberalism, following the breakdown of the (varying) combination of Keynesianism, class accommodation, public welfare provision and the acceptance of state intervention in order to maintain the lineaments of the settlement that had lasted in much of the industrialised capitalist world since the end of the Second World War. The eventual ascendancy of neoliberalism was not inevitable, and was indeed contested. As Harvey has it, 'The capitalist world stumbled towards neoliberalization as the answer through a series of gyrations and chaotic experiments that really only converged as a new orthodoxy with the articulation of what became known as the "Washington Consensus" in the 1990s' (2005, p. 13). Moreover some of the crucial battles took place in cities, not least in London. The head-to-head political confrontation in the 1980s, between the Thatcherites and the left-wing Greater London Council, over what would be the future of London was precisely between 'different visions for the city, which also turn out to be different visions for the wider society' (Cochrane, 2007, p. 145). It was a confrontation between two alternative ways out from the dissolution of the postwar social democratic settlement. And the victory of the 'neoliberal' vision for the city was to have effects throughout the country and indeed internationally. This was a class victory, though the nature and composition of the class was radically, and unevenly, reworked in the process. In other words, and as the analysis which follows indicates, 'the neoliberal turn is in some way and to some degree associated with the restoration or reconstruction of the power of economic elites' (Harvey, 2005, p. 19). The sharpening of inequality has everywhere, including even in the poorest countries, been primarily the result of the growth of a stratum of super-rich. As we shall see, this is certainly the case in the United

Kingdom, and within London specifically. Indeed, around
the world, as the poor have crowded into urban areas so too
have these places become the home-base of the new (and
old) wealthy. From here the economic orthodoxy is asserted,
and the vital importance of global cities is proclaimed in
countries that are ever more geographically unequal. A rhetoric of obeisance to 'the market' and the removal of the
potential questioning of market forces from the realm of
political debate (they are just 'natural') are central to this.[5]
So also, within the UK specifically and more widely, is the
power of the financial sector and the whole cultural and
institutional infrastructure that surrounds it. 'The City' and
its sectoral and social environs are central to the story of
what is happening in London and to the current exacerbated
reproduction of the North–South divide in the UK. It is
around its financial pre-eminence that London is most often
accorded world-city status. It is a story, replicated with varia-
tions in country after country, that has been repeated so
often and so loudly that it has come to have the status of
common sense. It is difficult to think otherwise. And yet, it
will be argued here, the repercussions must be questioned.
Within the global city the dynamics of this particular form
of growth produce poverty as well as wealth. In the nations
within which these cities are set, their growth comes at the
expense of other regions – but 'the regions' find it difficult to
object in the face of claims that the global city is the golden
goose without which all would be lost. And around the
planet as a whole, caught in the structures orchestrated from
the elites in these cities, the inequality deepens.

*   *   *

'Geography' is integral to all of this. Neoliberalism develops
unevenly. Cities are crucial to its functioning. New political
questions are raised about the intersection of territorialised

politics and wider responsibilities. Social and economic divides are exacerbated by geography. And so on. One small example: in August 2005 it was reported that:

> Directors' pay at Britain's top companies climbed an average of 16.1% last year – four times faster than average earnings and eight times the rate of inflation. (Finch and Treanor, 2005)

It was an increase that took average pay for a chief executive to more than £2.5 million. Nor was this a one-off – increases in previous years had been similar. In contrast, earnings for people in the country as a whole rose by 4.1 per cent to reach an average annual salary of £22,060. 'An average chief executive is paid 113 times more than an average UK worker' (ibid.).[6] This inequality of the extremes is characteristic of the 'Anglo-Saxon' version of neoliberalism (Hutton, 2002), and it has been growing.

It also has a geography. Such wealth is overwhelmingly concentrated, within the UK, in London and the South-East. An immediate effect, therefore, of such pay rises is a further widening of the national North–South divide. But there are other effects too. Mainly as a result of such vast incomes, London is the most unequal place in the country, and the effects of this wealth reverberate throughout the capital. For those Londoners who are already home-owners this may be a great boon, but the consequent constant rise in prices further fuels the gap between South and North. Meanwhile the effect on the poor in London is to make their poverty yet more marked as the cost of everything spirals out of reach. And so on. The geography of inequality becomes both consequence *and further cause* of national levels of inequality. The regional inequality between North and South is intimately tied up with national social structure. It is a constellation that overflows too, as we shall see, into issues of class and of democracy – into the maps of power more generally.

Moreover, in the political arguments over such issues, 'geography' is mobilised in conflicting ways. There are those (indeed there are many) who argue that the growth and wealth of London are of unalloyed benefit to the rest of the nation (just as the rich are sometimes argued to be of benefit to the poor). There are those who evoke an almost moral geography, imagining regions and countries as autonomous entities, 'competing', 'succeeding', and 'failing' economically. Much effort is put into constructing coherent place identities to cover over the conflicting interests that happen to share particular territories: a 'we' that binds together rich and poor against some other place elsewhere. There are the unequal binary geographies of self and other, us and them.

Lying behind the political mobilisation of such stories are more general geographical imaginations and implicit conceptualisations of space. They are understandings that are utterly essential, even if most often implicit, to our more general 'political cosmologies' (Fabian, 1983). For example, the notion that regions are coherent entities that compete against each other is part of a wider world view in which identities are autonomous, pre-formed before they come into relation with each other (Hudson, 2006a). This is the philosophy of the isolated individual, and the imaginary of a geography always already territorialised is a specifically spatial articulation of this view. It is an imaginary that has been fundamental to modernity and to the common-sense, as well as politically useful, assumption behind the division of the world into states and regions (Massey, 2005). An alternative geographical imagination would argue that the character of a region, or the economy of a place, is a product not only of internal interactions but also of relations with elsewhere. Not even islands are islands unto themselves. There is a constitutive interdependence. Space is relational. Such an apparently nuanced shift of emphasis radically challenges the judgemental imagination of places competing, succeeding and failing as a result of their own intrinsic characteristics alone.

Or again, and relatedly, that imaginary of place which focuses solely on its internal construction, and thereby obscures a potentially outward-looking local consciousness, identity and politics, fails to take account of those relations that run out from a place – that help construct its identity and on which it depends. In London, it focuses on the internal hybridity, or the triumphs of finance, without imagining the wider global geographies in which these are set. That view of 'the global' as simply an aggregation of 'locals' can be related, too, to a counterposition of local and global in which the local is revered as the hearth of authenticity, real lives, cultural richness, and so forth, while the global is imagined as some kind of place-less realm (a 'nowhere') which, by contrast, is powerful, inauthentic, somehow abstract. In such an imaginary the local is perpetually the victim of the global. Global forces arrive from 'elsewhere' and wreak havoc on a previous local embeddedness. And often – in place after place in the global South, in devastated mining and manufacturing areas of the global North – this is not an unrealistic reading. And in such a situation the response is often to defend local place. But the global is also locally produced; and global forces are just as material, and real, as is the local embeddedness. Some local places are the seat of global forces. And in such a situation it may be the local place itself – what it stands for, what its identity depends upon – that must be challenged.[7] Most places are complex combinations – on the receiving end of some wider forces, seat of the production of others – and in consequence in each case the political potential will be different.

Or yet again, in some imaginations of the geography of the world, 'space' is a surface, across which investments/migrants/connections flow and forces march. So those who would convince us that London is the golden goose of the national economy, whence flow benefits for the rest of the country, or that the financial City must be supported so that its growth can solve the problems of urban poverty within which it is set, or those who see 'globalisation' as flowing out across the world from its centres of power (as 'progress' once was said to do) are implicitly mobilising, most often, this imagination of space as a surface. It is indeed the necessary geographical correlate of the economic imaginary of trickle-down. Things flow outwards from centres to those who are the recipients. It is a colonial space, in which there is only one actor. The recipients are merely recipients. But if 'space' as a dimension is anything at all, it is the dimension of coexisting actors, the dimension that precisely enables (and requires) their multiplicity (Massey, 2005). So London, internally, is not a kind of pyramid, with finance as its shining citadel and the rest of us in one way or another dependent upon it. London is rather, as is space in general, a field of multiple actors, trajectories, stories with their own energies – which may mingle in harmony, collide, even annihilate each other. So the benefits of London which are imagined to flow out to the grateful regions of the North, say, in fact intersect in that North with other stories, with the existing economies and cultures of those other places, with repercussions that in each case must be analysed anew.

That recognition of multiple trajectories is one implication of taking space seriously (Massey, 2005). It forces respect for the coeval. Many political cosmologies, in contrast, are framed in such a way that 'others' (other actors, other trajectories) are in one way or another either obscured from view or relegated to some sort of minority or inferior status. Such cosmologies refuse the challenge of space. The smoothing of space into a surface, as just described, renders invisible the other actors on the stage. It allows certain political questions – as we shall see – to go unasked. Equally, there is that political cosmology that is framed in such a way that other actors are relegated to a historical past (they are developing, we are developed). Here, the contemporaneous multiplicities of space are denied, and 'history' is reduced to the singular linearity of 'there is no alternative'. In this geographical imagination the 'others' are demeaned, their actually existing difference realigned to being merely a matter of their being 'behind' in the historical queue. This is an understanding of space (in truth, a turning of space into time, geography into history) that refuses coevalness.[8]

Yet space is the dimension of contemporaneous existence. It is, if taken seriously, the dimension that demands an attitude of 'respect' (Derrida, 1997). Ken Livingstone's stress on that term – on that stance in relation to others – was therefore significant. Deploying implicit conceptualisations of space which refuse that recognition of coevalness, and thereby evade the challenge of respect, is a fully political manoeuvre. It feeds yet again into what will be a recurring theme in the arguments that follow: that the geography of the world is intimately entwined with the most fundamental of political issues: with inequality, with recognition and the evasion of it, with class and with democracy, with – what we inevitably live within and are constantly remaking – maps of power. Just as we make history in retrospect, through the stories that are told, so also we make geographies, through the implicit imaginations we deploy. And geographies relate not only to the past but also to the present.

Geographical imaginations are most often implicit. When deftly mobilised they pass us by as self-evident, barely recognised as being the framing assumptions that they are. And in part precisely as a result of this, they are powerful elements in the armoury of legitimation of political strategies. They are part of the scaffolding from which hang grand political visions; they are integral to understandings of such things as globalisation and development; they are embedded too in the assumptions that underpin the most apparently mundane of government documents, or the research reports of private hired consultants; and they weave their way through the street-level practice of political activism. As we shall see, they are essential to the story of global cities – for instance to those who argue for the inevitability and necessity of such cities. And challenging these hegemonic imaginations is therefore essential to any wider political challenge. It is important moreover that not only do we expose the hegemonic geographical imaginations (as in Said, 1985; or Gregory, 2004) but that we also take the further political step of proposing alternatives. In most situations there are multiple geographical imaginations in play. And even though it may rarely be evident, there is contest between these imaginations: 'whose geography?' Very often the politics of a situation would be clarified were the geographical imaginations and the conceptualisations of space made explicit. That is one of the aims here.

*   *   *

This book grew out of the weaving together of two long-held preoccupations. On the one hand, there is my interest in the politics both of regional inequality in the UK and of London in particular. My earliest geographical research was in industrial location and issues of regional uneven development. Subsequently, as a Mancunian facing the pressures of the professional job market, especially for women, I moved south. By the 1980s I was involved in London politics as a board member of the Greater London Enterprise Board

(GLEB), the economic policy arm of the Greater London Council of that decade, famously abolished by Margaret Thatcher in 1986. And I have remained in touch, in a variety of ways, with London politics both official and grass-roots, ever since. On the other hand, the book grows out of more general concern with the politics of geography and the conceptualisation of space, and with the challenges that space throws up (Massey, 2005). These two preoccupations have never been separate, the one apparently 'applied', the other 'theoretical'. Nor was the movement between these preoccupations just in one direction (from the theoretical to the applied, in those linear modellings of knowledge as though it ran only one way, from 'theory' to 'practice'). The most startling disruptions to 'theory' have often come in the midst of engagement with particular, political, issues. I still have the little books full of 'theoretical' thoughts I jotted down on the tube coming back from meetings of GLEB. London at the beginning of the twenty-first century, and in a world in which the neoliberal hegemony seems more assured than it did in the 1980s, faces different issues, and poses politico-theoretical challenges that are new. They are also issues and challenges which in more general terms confront, or should confront, any city claiming, as so many do, to be global.

Part I

INVENTING
A WORLD CITY

1

CAPITAL DELIGHT

There is a buzz about London, as about so many big cities. As well as draining you, utterly, as you battle through crowded tubes and buses and grimly negotiate the hubbub, it returns that lost energy to you. Virginia Woolf's evocations are as true as ever. And London is enriched now with an increased cultural diversity, and a sense that the city is going places. (And having 'our own' government again reinforces this.) There are shared understandings of how to behave, the mutual acknowledgements that are barely acknowledgements as we negotiate around and past each other in the flood. There is geographical variety, for London is also a constellation of local centres. Even the disputed skyscrapers and the maligned Canary Wharf give a thrill. I was at times involved in the battles against these latter developments (the class wars over what would be the future of the Isle of Dogs). But it was with a kind of guilty delight that with a friend, a visitor from 'the North', I rode round and round some years ago on the Docklands Light Railway. With a day-pass you can treat it like a funfair, its tinny precariousness making it feel even more like a ride in a fairground in and amongst the amazing concrete and glass. The ambiguities abound: I *campaigned* against this stuff! Yet even as we object, the very energy, the preposterousness of it all, enthralls. (I think of my home city – Manchester – now glowing again in its cleaned red brick, the great cliffs of buildings lining the central streets. We love them now, but they were once the commercial bases of imperial cotton and a working class in clogs. Will we come to love Canary Wharf – brash nodal point in an equally imperial new world economy – just as well?) And there is Westminster, and the walk along the South Bank, the great wheel of the London Eye, and Tate Modern, converted from power station to art and from where you can look across the river at St Paul's and the City. God and Mammon. There is the expansive openness of the parks, and the endless passage of aircraft overhead. This is a city it is very easy to celebrate.

We have become accustomed to this dynamic, bustling London. Yet only a few decades ago the dominant images were different, as were the issues that headed the political agenda. Then, it was inner-city decline, and the collapse of manufacturing and of dockwork, that were the focus of economic attention. In the mid-1980s the radical Greater London Council, then ensconced in County Hall, famously displayed from its roof the city's unemployment numbers, taunting the Houses of Parliament across the river, where Margaret Thatcher reigned supreme.

She reigned over a country deeply divided by geography, over a disappearing landscape that was utterly different from that of today: a nation of regions dominated by mining and manufacturing, in which trades unions and trades councils had serious social bases; a nation in which manufacturing regions were 'global' regions (Hudson, 2006b). It was the final crumbling of an imperial geography. The national spatial division of labour, the internal employment geography of a country, exists in intimate relation with the role of that national economy in the world beyond, its place in the international division of labour. Britain's international role as imperial and colonial power had long been fading, its future was up for question. The same was true, in consequence, of the national geography. But while among the working class in the North and the West, and in London's docklands, the decline of the production and material trading bases of Empire continued to create poverty and unemployment, the financial side of that imperial role in its tightly knit class cohorts, operating from London, had woven a basis of social relations and international lines of authority upon which it might reinvent itself.

In the 1980s, however, the future still seemed so open. On the one hand there was defence against the destruction. In the coal regions the miners came out on strike. In the east of London, the battles over what was to become of the docklands figures now as a condensation of the terms of the struggle over the future of the capital city. On the other hand there were also the beginnings of the formulation of a different, alternative, future. In the cities a new urban left began to do battle against the domineering new orthodoxy. The GLC itself was working to formulate a way forward that was neither a simple defence of the old nor a capitulation to the emerging neoliberalism of Margaret Thatcher (see especially GLC, 1985; Mackintosh and Wainwright, 1987). At the very moment when it seemed, so briefly, that these political wings, of resistance and of an alternative future, might seriously join forces, each was defeated – the cities amid recriminations occasioned by the brutality of the personalisation of liability for ratecapping, the miners by a concerted onslaught of fair means and foul, the GLC (along with the other – also radical – metropolitan councils) quite simply abolished.

It is out of this that has emerged the London of today. As the Greater London Authority reported in 2004, 'London has reinvented itself' (GLA, 2004a, p. 13). The regrowth is registered even in the simple terms of population numbers: from a peak in 1939 of 8.6 million, the population of Greater London had fallen to 6.8 million in 1981, but began to rise again in the late 1980s to reach an estimated 7.3 million in 2003 (Hamnett and Randolph, 1982; GLA, 2004b).

The usual account of this reinvention describes it in terms of a change in balance between sectors of the economy and revolves around that shift from dockwork and blue-collar manufacturing towards finance and business services that was at the heart of the battles over the East End. Between 1978 and 2000 finance and business services grew by 81 per cent in Greater London (582,000 jobs) while manufacturing declined by 63 per cent (432,000 jobs). The trends were similar in the Outer Metropolitan Area (166 per cent and 385,000 jobs in contrast to -48 per cent and a loss of 283,000 jobs) (Buck et al., 2002, p. 97). And, in social terms, even these figures underestimate the change, for, as Banks and Scanlon (2000) document, possibly over half of those recorded as employed in 'manufacturing' in London work in the head offices of manufacturing companies that 'produce' elsewhere (cited in Hamnett, 2003), and that too is a characteristic that has been reinforced over the same decades. Nor were such shifts unique to this city. Fainstein and Harloe were referring to both London and New York when they wrote of the 1980s:

> While other economic sectors were dispersing geographically, certain advanced service industries centring around financial activities intensified their presence in the centres of these cities. This intensification resulted primarily from the enlarged role of financial capital in co-ordinating the

world economy and the extremely active deal making that accompanied this role. The decade witnessed the birth of new financial markets for the exchange of arcane financial instruments and the raising of huge pools of credit to underwrite speculative activities in property development, mergers, and leveraged buy-outs, as firms themselves became negotiable assets. (2000, p. 155)

However, to talk of sectoral shifts is to register only some particular, measurable, outcomes of more complex processes. As Fainstein and Harloe indicate, what was under way was a shift in the world economy – the beginnings of the assertion of what has come to be called neoliberalism. And integral to that shift was its spatial reorganisation, an essential element of which, in turn, was the emergence of what came to be called world (or global) cities.[1] Nomenclature does not, however, necessarily indicate rigorous definition, and there has been much debate over what it is that makes a place a world city. Beaverstock, Smith and Taylor make an initial distinction between 'a demographic tradition which is largely interested in the sizes of cities and a functional tradition which treats cities as part of a larger system' (1999, p. 445). As they say, there may well be overlap between these classes of cities: 'New York and Mexico City, for instance, are both mega- and world cities' (ibid.), but it is nonetheless a crucial distinction: 'Calcutta is a mega-city but not a world city, Zurich is a world city but not a mega-city' (ibid.). It is also a distinction that evokes again the stark contrast between some cities of command and some cities of the majority world (and Davis's cities of slums).[2]

Even within the functional tradition, there are different possible sets of criteria. Beaverstock et al. adopt a two-pronged approach, both scanning the whole range of definitions and rankings in the literature and constructing their own ranking by following Sassen's argument 'that it is advanced producer services which are the distinctive feature of contemporary world city formation' (1991, p. 446). (It is worth pausing for a moment here, to consider this adjective – 'advanced'. It is also deployed by Fainstein and Harloe, and is indeed in general use. But the effect of its mobilisation is normative – it implies an inevitability of direction; it naturalises what in fact happened, implying that there was no other way in which the economy might have developed. 'New' might be a more neutral term.) Anyway, what this 'advancedness' is taken to denote in this context is banking, accountancy, law and advertising. Through their two-pronged approach, Beaverstock et al. demonstrate an exact coincidence between the four cities identified by all the sources in their review and the four cities with maximum scores in their own inventory. These cities are: London, Paris, New York and Tokyo.

It is difficult to dispute, in its own terms, this characterisation. A *Financial Times* report in 2006 on *The new City* revelled in its dynamism, inventiveness and success: 'There is a certain swagger about the City of London these days', writes Martin Dickson in its opening article (Dickson, 2006). The City of London's website begins its 'Key facts about the City and Corporation': 'The City of London is the world's leading international financial and business centre – a global powerhouse at the heart of the UK's financial services.' It goes on to document some relevant information: that in London each day takes place over 30 per cent of global foreign-exchange turnover, over 40 per cent of the global foreign equity market, 70 per cent of all eurobond trading. There are markets in international insurance, international futures, derivatives, metals. There is the management of pension funds, the arrangement of international bank lending, and there is the maritime industry.

There are over 250 foreign banks in London, and over 550 foreign companies listed on the London Stock Exchange. Three quarters of the Fortune 500 companies have London offices (City of London website, accessed 22 November 2006). And around this financial core there is a vast and intricate constellation of other business services, expanding in employment terms over recent decades even faster than is finance.

London reinvented itself, then, as a global city. But this too is a formulation worth pausing over. It is, for one thing, a quite different evocation of world citydom from that notion of 'the whole world in one city' (Livingstone) and 'Here lives the great music of humanity' (Okri) that was appealed to in the aftermath of the bombing. The two different narratives of London-world-city sometimes run in parallel and sometimes intersect – a pattern that will be investigated in what follows. However, even this notion of global citydom based on finance and business has itself been challenged. There are a number of reasons for this. It has been challenged for a Euro-American bias (King, 1990, 2000; Robinson, 2002; Olds and Yeung, 2004; Godfrey and Zhou, 1999) which sets up certain Western cities (including London) as norms against which others then come to be judged. It mobilises universalist assumptions that obscure the fact of situatedness. Its rhetorical force, what Douglass, reflecting on Pacific Asia, calls '"world city" as the new shibboleth of global achievement' (1998, p. 111), pressures urban governments and managers around the world into striving for this self-same thing. And it dooms to invisibility, to 'structural irrelevance', cities which do not rank on its hierarchial league-tabling of the planet (Robinson, 2002). Robinson goes on to argue that the concept itself, along with the panoply of categories that accompany it, should be abandoned in favour of a recognition of the specificity of each and every city. All of them are 'ordinary cities' (see also Amin and Graham, 1997). This is a strategy that chimes well with the argument in this book – we need to value and build on the *diversity* between cities – and it is strongly reinforced by other criticisms of the notion of global cities.

For instance, this a characterisation of world citydom that prioritises place in the world economy and that makes a claim for dominance, in terms of command and control, within that world economy, of finance and business services. This too is a characterisation that has been challenged. Thus Smith questions 'the common assumption that the power of financial capital is necessarily paramount', and 'the criteria according to which cities come to be dubbed "global". If there is any truth to the argument that so-called globalization results in the first place from the globalization of production, then our assessment of what constitutes a global city should presumably reflect that claim' (2002, p. 437). Or again it could be argued that cities can be world cities in all kinds of other ways too, as dominant foci in particular spheres of activity: Jerusalem as a world city in the sphere of religion, Sydney perhaps in lesbian and gay networks, Hollywood and Mumbai for film. Indeed, as we have already seen, London can itself claim to be a world city in more than one way. That the unqualified denominator 'world city' is usually applied to financial and service centres is precisely a sign of the material and discursive dominance of a particular reading of what Sassen calls 'the current phase of the world economy' (1991, p. 126). Were such debates to be taken more seriously, one small effect might be, through the recognition of what else is happening in the world, to reduce just a little that 'certain swagger' of London-global-city.

Let us, for a moment though, take this global citydom on its own terms. To define it in this way is, then, to make reference to an aspect of its current role in the contemporary world economy. London is undoubtedly a significant centre of coordination of that economy. Indeed, it is *on the basis of establishing this role* that the city reinvented itself after its mid-twentieth-century decline. It was this future that won out over all alternatives, from the understandable but perhaps unrealisable desire to cling on to or even rebuild an economy of dockwork and manufacturing to the more radical and politically challenging possibilities hinted at by the GLC.

Even here, however, there can be different emphases. Hamnett, for example, emphasises the classic aspect of internationalisation as the foundation of London's new surge of growth:

> in order to understand the basis of these changes, we need to understand London's changing role in the global economy and financial system. During the 1970s, 1980s and 1990s London has strengthened its role as one of the major control centres for the global economic and financial system. It functions as one of the major 'world cities' (Friedmann, 1986) or 'global cities' as Sassen (1991) prefers, and exercises a leading role in the organisation and control of the world's economy, trade and financial flows. (2003, pp. 4–5)

Buck et al., in contrast, emphasise deregulation:

> Though some boosts to London's growth have come from seemingly unrelated sources (notably from new flows of international migrants, . . .), there is clearly some connection between the city's current affluence, productivity and dynamism and the great liberalization of markets that occurred in the 1980s – as well as some connection with internationalization. (2002, p. 135)

There is a battle here over representation. The different geographical imaginaries in play can be used to support distinct political arguments. As Gordon writes, 'different representations of a city's role can be mobilised in support of alternative political strategies and claims for resources' (2004, p. 4). In a number of forensic analyses, he has disaggregated and refined these claims.[3] 'Currently', writes Gordon, 'the fashion is to characterise London as, first and foremost, a "global city". Sometimes this means that its population and workforce have become (particularly since the late 1980s) remarkably cosmopolitan, by most international as well as British standards. More often, however, "global" refers to the significance of transnational command, control, financing and service-providing roles which it undertakes within an increasingly integrated world economy' (ibid.). However, he argues, this characterisation needs investigating. It is certainly true that these are London's most *distinctive* functions, distinguishing it not only from other cities in the UK but also from other global cities.[4] But it is not these global command and control functions that have been 'the main driver underlying the city's growth since the 1980s' (ibid., p. 5). Nor do these functions represent the major element of London's export base.[5] Rather, London's main export market is in fact the 'rest of the UK'. This is true for the London economy as a whole, from which the rest of the UK (RUK) takes 28.5 per cent of all exports, compared with 12.33 per cent going abroad. Moreover it is equally the case for those central activities of London's world citydom: financial services and business services. For financial services, the comparable percentages are RUK 39.88 per cent and international 31.46 per cent and, for business services, RUK 32.89 per cent and international 12.08 per cent. Gordon adds the nice observation that 'On an income basis, the "global" share would be rather higher, since workers in the "City" finance centre typically earn half as much again as the London average – but the point about the dominance of the national market still remains' (ibid.). As Gordon concludes:

> Wider processes of globalisation – including the financial and labour market deregulation which it has provoked – have clearly had pervasive, and *economically* favourable, effects on London since the early 1980s, . . . But the narrower process of 'globalcityisation' does not seem to have contributed disproportionately to this shift, . . . Even within the City's 'wholesale' financial services, rising domestic demand has been a factor of comparable importance, over a period when people have been encouraged/required to take much greater financial responsibility for their own housing, pensions, health care and education. (Ibid., emphasis in the original)[6]

It is important to be clear about the argument here. *First*, this analysis gives the lie to any notion that London, in economic terms, is floating free from the rest of the UK economy into an international arena of its own. This is plainly not the case. This directly contradicts the conclusion of Dorling and Thomas (2004) that in a globalised economy London does not need the markets of northern Britain. As a London School of Economics study puts it, 'the London economy is still closely integrated with the overall UK economy' (LSE, 2004, p. 103). This is an important point to establish because it leads on to a further set of questions – about what, then, is the nature of that integration (on which regions' terms? to which regions' benefit?) – which will be the focus of Part II.

Second, however, and to return to the characterisation of London itself, none of this means that London is not a global centre of command, playing a crucial role in framing the world economy in neoliberal form. That is agreed by all. It is also the case that finance and business services have been of crucial significance in the resurgence of London in recent decades. What is important is that, *third*, the constant repetition of the importance of London's global export role is also a *performance* for internal, national, consumption. It is a claim, and a legitimation. It is mobilised, for instance, to justify the untouchability of London's financial and business sectors. It is mobilised, as we shall see, in the constant battles for resources. And yet, as we have seen, the specifically international role of these sectors is not overwhelming. It is not the basis of London's reinvention and resurgence. Rather, *fourth*, it is *deregulation*, or more generally what has been called the project of neoliberalisation, both internationally and nationally, on which London has fed itself (see also Dickson, 2006), including in particular the colonisation by private capital of industries and services formerly provided by government (the utilities under Thatcher and Major, significant parts of the welfare state, especially health and education, under Blair). It is a project that has entailed also the promotion of competitive individualism and individual self-reliance (as 'people have been encouraged/required to take much greater financial responsibility for their own housing, pensions, health care and education' – see Gordon, 2004, p. 5), and that abandons previous notions of mutuality and systematically undermines the idea of the public good. It is, in other words, more than a question of economics; it set the scene for a shift in a whole way of being. It is all this that has been the foundation of London's reinvention.

But this group of sectors is still only one part of London's economy. Indeed, one of the characteristics of cities generally is often held to be the variety which underlies their creativity and dynamism, that juxtaposition of differences that can produce something new (Amin, Massey and Thrift,

2000). This is certainly true of London. What emerges most emphatically from any detailed examination of London's economy is its sheer diversity (Buck et al., 2002). Gordon characterises London's economy as 'extraordinarily diverse' and argues that 'Complexity is the essential characteristic of the London economy, and it is this rather than any specific segment of demand which deserves to be seen as its key driver' (2004, p. 7). There is a *politics* then (in the broadest sense of that term) in this continual characterisation of London as overwhelmingly global and in the highlighting of this particular element (global finance and business services) of its complex, diverse, economy. It serves a purpose.

This manoeuvre of highlighting only one part – and often only a small part – of the urban economy is typical of global-city discourse (see, on this, Amin and Graham, 1997; Amin, Massey and Thrift, 2000; Robinson, 2002). It is a strategy of synecdoche, where the part is made to stand in for the whole. So London and New York are classified as global cities on the basis of their finance and associated industries; but that is a characterisation that obscures all the other vital elements of their economies and societies. (Equally, the characterisation of some cities in the global South as simply cities of favellas and squatter settlements ignores the other components of these huge metropoles – Robinson, 2002.) In a similar way, the division between 'basic' and 'service' sectors of an economy (see note 5) highlights the sectors that are distinctive (basic) and relegates to comparative invisibility the host of activities that keep the economy going – and which likely take up the greater part of its activity and provide the greater part of its employment. It has to be said that this is a tendency not confined to such issues of urban economics. The wide literature that deals with identity more generally, and especially that part of the literature which draws on psychoanalysis, frequently insists on defining identity as differentiation, as difference *from*; its stress is on the constitutive outside. In this kind of approach the only thing that can stand as an identity, of a city, an individual or a nation, is what differentiates it from what it is not. It is a mode of identification precisely designed to ignore what we have in common. On such a criterion, global finance is indeed London's identity (and the above critique of global-city discourse fails) because these are (among) its most distinctive features. But if what is at issue is the character of London, what makes it tick, what keeps it going, what it is to live here, then it is necessary to include far more.

This strategy of synecdoche, however, serves political purposes. Hamnett separates politics from economics, seeing the two spheres as alternative explanations of London's transformation. This can be a dangerous argument, and indeed the common acceptance of the separation is itself a political achievement, for it can naturalise 'the economic', removing it to a realm of inexorability supposedly beyond human intervention.[7] More significantly, the fact that London was reinvented *in this way* was the outcome of a political contest. This was a class victory, in part secured by the establishment as hegemonic of a particular socio-economic doctrine – neoliberalism. And the cry of 'London-global-city', as we shall see, helped to secure that victory; it was a means of legitimating one side in the political contest, at both national and city levels, over what would be the future direction of the economy after the breakdown of the postwar settlement. Most obviously, and for all the Tory talk of 'national sovereignty', the first thing Margaret Thatcher did on coming to power in 1979 was lift restrictions on foreign currency exchange, to be followed in the mid-1980s by the deregulation of the City (the so-called Big Bang). Moreover these moves were set in the wider context of a whole gamut of deregulatory and commercialisation policies, in pensions, housing, health care and education, which considerably increased the market for City activities (it is these that Gordon is referring to in the quotation above). Indeed, as Hamnett records: 'Thatcherite policies, including deregulation, privatisation, interest rates, and higher-rate income tax cuts, benefited the private sector, financial services, the middle classes, London and the South East at the expense of the public sector, manufacturing, the old industrial regions and the working classes' (2003, p. 16; see also Hudson and Williams, 1995; and Peck and Tickell, 1992). On the other side of this contest, in London itself, 'the GLC leadership was engaged in a political project not simply to rebuild the base of the London Labour Party, historically dependent on skilled manual workers who were now disappearing from the city, but also to demonstrate on a small scale the potential of an Alternative Economic Strategy which could be pursued nationally by a Labour government' (Gordon, 2004, p. 3). In fact, the national Labour Party was as hostile to the democratic radicalism of the GLC as were the Conservatives, and did little to oppose its abolition, thereby leaving the field clear for Thatcherite neoliberalism (and leaving itself in a policy vacuum from which would emerge the 'Third Way').

London, then, rose to growth again on a tide of neoliberalisation. Its resurgence is a product of (a particular form of) deregulation and privatisation/commercialisation, along with internationalisation – what the LSE study, produced for the Corporation of London, calls 'an economic environment more favourable to London's strengths' (2004, p. 18). It *benefited*, in other words, from 'the current phase of the world economy', and to that extent its claims of 'success' – as though it were due to some kind of superiority of moral fibre over other regions of the country – are spurious (see Part II). London's 'success' over recent decades has been due to its inherited position within wider structures. Moreover in that inherited position, London's existing 'strengths' were a direct product of its previous role as a central imperial and colonial city. As King writes, 'the phenomenon of colonialism' was 'the mode in which London became incorporated into the global economy' (1990, p. 10); that dominant role in an older international division of labour was crucial in laying down 'the infrastructure, whether . . . economic, political, social, cultural [or] physical and spatial, for the later world city' (ibid., p. 46).

Nonetheless, this particular form of reinvention has been due also to the ability of these parts of London's complex and diverse socio-economic structure to take advantage of that inheritance, to conjure a new geography for a new round of investment. Thatcherism, and the New Labourism that followed it, were – and are – strategies that play to those very interests. And it was within that context, from the development of the eurodollar market in the 1960s, that London's financial sector took the lead in the development and promulgation of the deregulated international economy within which it now holds such a prominent position (Christensen, 2007). In other words, it is not only that certain parts of the London economy function, now, as a command and control centre of the reworked international economy, it is also that the City was in at the beginning, inventing and taking the lead in developing that very reworking. There was, moreover, a global geographical repositioning going on here too. On the one hand, this self-reinvention has entailed a subordination to, and mimicking of, the USA. It is often said that imperial decline has led to a skewing of British geography away from an Atlantic orientation and towards Europe (the contrast between the decline of the port of Liverpool and the digging of the Channel Tunnel is frequently called in evidence). However, in the matter of the financial heart of London, certain aspects of reorientation have been towards the USA. Indeed, as Hutton (2002) details at length, domination of UK economic thinking more generally by US influences has been marked. It is this that is sometimes termed the Anglo-Saxon model of capitalism.[8] As Hutton has it, finance has been an area where 'Britain has been the willing object of the American embrace':

> From Britain's failed attempt to return [to] the gold standard in 1925 to the systemic trade deficits of the immediate postwar era, it looked as though the City of London would have to cede its role as an international financial centre to New York because of the weakness of sterling and the strength of the dollar. But from the mid-1960s the City took advantage of an offshore status manufactured by British taxation policy to create a market in dollars owned by non-Americans (eurodollars), the supply of which was assured by the ambitions of American multinationals and America's trade deficits. London, in short, became an offshore extension of New York, creating a major market in eurodollars which now makes it the world's biggest international financial centre. (2002, p. 36)

It has been a lucrative subservience, for some. It is out of this that the new elite has been born.

It is centred on finance, on the City. However, the establishment of this new position involved huge disruption, both social and spatial, even within the financial sector itself. Big Bang, in October 1986, opened up the old settled ways. 'Big Bang turned London's equity market from a protected, cosy old club into the vibrant global centre it is today' (Marks, 2006). The old social exclusivity, the intricate protective practices, even the leisurely working hours, were abandoned (though not the predominance of white males). On 27 October 1986, 'Electronic trading systems were switched on

and the secluded cabal of privately owned firms that dominated City share dealing was smashed' (Treanor, 2006). All
this was necessary to establish a new form of global financial
dominance. It was a price worth paying.

Yet, the consolidation of this new position has also
involved taking advantage of, and reworking, characteristics
inherited from the previous incarnation. The symbols of the
old Empire, often icons of the very characteristics that had
now to be abandoned, were retained. If the old class ways
had, perforce, to be opened up, nonetheless they could be
mined to shore up the legitimacy and the social cachet of the
new. As Jacobs has it, 'the traces of imperial might continue
to condition the ways in which cities like London reorient
themselves within new global and regional arrangements'
(1996, p. 24). Her analysis of planning battles in the City
demonstrates clearly that 'Imperial nostalgias are not tied
just to preservationist interests' (ibid., p. 40). Thrift, though
making no mention of the imperial history from which it
derives, refers to a 'heritage style', and emphasises its significance in solidifying London as a centre of 'cultural authority': 'The old gentlemanly discourse may have dissolved but
the "trappings of trust" still remain: quiet, wood-panelled
dining rooms, crested china, discount round top hats, City
police uniforms and so on are all used to "brand" the City, to
boost its image of solidity and trustworthiness' (1994, p. 350).
The Corporation of London in its 1986 plan for the City
opined that 'The City of London . . . is noted for its business
expertise, its wealth of history and its special architectural
heritage. . . . The City's ambience is much valued and distinguishes it from other international business centres' (Corporation of London, 1986, p. 3, cited in Jacobs, 1996, p. 55).
Thrift cites Pugh: 'The rediscovery of tradition is the key to
City trendsetting' (1989, p. 127). That tradition is imperial.
The same is true of spatial form. The old, tight, regulated,

spatial concentration of the functions of the Square Mile has, even in its explosive expansion, proved a vital asset in the construction and maintenance of the discursive community that is now so powerful within the new, neoliberal, international division of labour (Pryke, 1991, 1994). Thus are elements of an old international division of labour cannibalised in the process of insertion into the new (Massey, [1984] 1995).

The emergence of the new elite, moreover, involved more than simply finance. The LSE study, for instance, stresses 'a far wider group of business services', in particular real estate, renting and business activities other than finance strictly defined (2004, p. 103). Beaverstock, Smith and Taylor's (1999) four criteria consisted of accountancy, advertising and law as well as banking/finance. Advertising, research and development, accounting, auditing and taxation, legal services, market research and consultancy, personnel recruitment, renting of machinery and technical consulting, investigation and security . . . all these and many more have grown rapidly as part of London-global-city. The CEOs in a host of corporate sectors, including the upper echelons of 'new technology', are part of this new elite. It is a vast and intricate constellation of interrelated activities – 'the service professions of capital' (Rustin, 2006) whose numbers, confidence and remuneration have boomed while other professions have lagged. Perkin writes of 'a running quarrel . . . between the two wings of professionalism, the public sector professionals, who wished to see an expansion of their services for everyone, and the private sector professionals, chiefly the managers of big business and their friends, who wanted less public spending and lower taxes. . . . The triumph of the Thatcherites . . . was the victory of the private sector professionals over the rest of society' (1996, pp. xiii-xiv). To these service professionals of capital are now added those who have more personally thrived upon the marketisation of the public sector 'emerging at the interface between government and the corporate sector, in an ambiguous and occluded area through which a great deal of public wealth is being transferred to the private domain' (Rustin, 2006, p. 9). Added to this again there is the land and property capital that oils the wheels, opens up new frontiers, clears the ground (sometimes quite literally). It is these land and property interests that have been a vital mechanism in the concentration of investment in commercial building in the prime central areas of cities and into the central cities such as London (Edwards, 2002). They have been crucial in the expansion of London's old financial centre out into the dense mixed-use areas that have for so long surrounded it. They have profited handsomely from the wider regrowth of the metropolitan area under conditions of inadequate land supply – itself a result in part of historic planning decisions and in part of that very growth itself. Moreover all this has taken place in the context of a long-term shift in the nature of landownership, away from what might be called 'industrial landownership' (where land is owned essentially as a condition of other production) and towards 'financial landownership' where the ownership of land is itself the means of extracting a profit (Massey and Catalano, 1978; Knox, 1993).[9] And the constellation includes too all those involved in the business of establishing and coordinating: the conventions, the rules, the discursive communities, the constantly evolving understandings. It is this whole institutional infrastructure that is crucial (see, for instance, Cohen, 1981; Daniels, 1991; Thrift, 1987). Such constellations are 'centres of interpretation' and, as Thrift (1994) argues, London's City increasingly markets itself as such, as a 'centre of cultural authority' for global financial services. It is moreover a centre of authority with effects far beyond the financial services. Financial markets and the constellation of activities that surround them are 'world making' (Pryke, 2005). And they are so in two senses. On the one hand: 'Places such as New York and the City of London are where time-space [it could be a soy field in Argentina, or a forest in Indonesia] is drawn-in, mathematized and made financial' (ibid., p. 4). On the other hand, the logics and conventions of these powerful discursive communities become woven into and gradually dominating of the hegemonic 'common sense' (Hall, 2003). As Sassen (1991, 2001) has long argued, it is not only a question of *functioning as* a command point for the global economy; it is also one of *producing the means by which* that commanding can be made effective.

This reinvigoration of London, then, is by no means just a matter of a statistical shift in balance between sectors of the economy. Rather it represents the rise of a new elite, and the culture in which it is embedded, and the victory of a new economics. It is the reassertion, in a remoulded form, of class power.

That 'certain swagger' referred to earlier characterises a whole group. Says Dickson (2006) of the new City: 'the swagger is visible in the people'. This is a group that knows it has won. The week before the celebratory *Financial Times* report appeared, the Chancellor's Budget had announced a new strategy 'to promote London as the world's leading international centre for financial and business services', which, as Daneshkhu and Giles dryly observe, 'came at a time when the City seemed to be doing very well by itself' (2006). Williams, writing of the City in its wider constellation, argues that:

in their various political, professional, financial and business forms, [they] were able to increase their authority by stages during the late twentieth century so that they have now grown to constitute the highly effective and seemingly permanent governing class of early twenty-first-century Britain. This has been a period of elite consolidation for which there is no parallel in the country's history. (2006a, p. 217)

Adonis and Pollard, likewise, argue that:

The rise of the Super Class . . . is a seminal development in modern Britain, as critical as the rise of the gentry before the English Civil War and the rise of organized labour a century ago, and rivalled in contemporary significance only by the disintegration of the manual working class. (1998, p. 67, cited in Lansley, 2006, p. 138)

This is a 'new class' in which the financial constellation is dominant and which, ultimately, traces the roots of its new wealth to deregulation. Williams writes of the consequences of 'the City hegemony over British life' (2006b), consequences that are political and cultural as well as financial and commercial. As he points out, there are few competing voices and, while the manufacturing bases and places of Britain's former empire are crumbling: 'This new financial elite is the true heir to the imperial legacy' . . . 'here is an elite of the elites whose power has grown to a dimension that is truly imperial in the modern world' (ibid.).

The rise of this group has been an important element in the widening national economic inequality of the last thirty years (starting with the huge leap in the crucial 1980s) (see, for instance, Institute for Fiscal Studies, 2006). Moreover, it happened at the same time as, and partly because it has itself provoked, a deadening political silence about, precisely, issues of class. As Lawson puts it, Blairism has inverted the politics of social democracy – that is, it has forced people to fit in with the demands of the market rather than moulding the market for people. In that, he argues, it is accepting a politics that is driven by a global elite composed of 'the new untouchables. Because of them, we cannot not talk about spiralling executive pay, rewards for failure, or wealth beyond imagination that allows some to spray champagne around West End bars for the conspicuous fun of it' (Lawson, 2006).

This London honeypot has also attracted the rich, and those wanting to be rich, from around the planet. The already rich come for 'tax reasons', and for the presence of that 'economic sector' of advisors who help manage the assets of the seriously wealthy. As Meek (2006, p. 6) puts it 'London [is] the destination of choice for the world's multi-billionaires. For the ultra-rich few, this country is now a virtual tax haven, which is why more and more princes, tycoons and oligarchs are making it their home.' Others are attracted by the lucrative opportunities in the City – 'more than one in 10 professional staff in the City of London come from countries outside the EU and the US' (Batchelor and Larsen, 2006). Here have come the plunderers of Eastern Europe and the old Soviet Union (Lansley, 2006, pp. 210–11). A report on French people working in the UK found 69 per cent of them in London and half of those working in financial services in the City. As *The Guardian* put it in a perceptive headline: 'Young exiles embrace the Anglo model' (Seager and Balakrishnan, 2006) – the 'exile', of course, is from the comparative egalitarianism of 'old' continental Europe. This rush to the trough is, then, another part of London as 'multicultural' world city – people come also for the feeding frenzy.

It is here, in this reinvented metropolis, that are based those directors referred to in the Introduction who in 2005, awarding each other their increments, were paid 113 times more than the average UK worker. This, too, links back to wider geographies:

> The pattern of British chief executives' pay is now openly modelled on the American lead . . . Over the last five years the average salary of a chief executive in Britain's leading companies, including bonuses, has more than doubled . . . (excluding share options), following the trajectory of the growth of American remuneration and bearing increasingly little relationship to company performance. Britain is importing American conceptions of inequality wholesale . . . (Hutton, 2002, pp. 37–8; see also Lansley, 2006)

In any case, it has resulted in levels of inequality far higher than in the major economies of continental Europe. This 'strong reinforcement of the position of the richest' in the context of more generally increasing inequality is character-istic of the Anglo-American model. It is most fiercely marked in the USA, but not found in France. The UK is fol-lowing the US model (Dunford, 2005, p.158). In the midst of campaigns against such salaries, waged both by sharehold-ers and by trades unions, this importation of US ways became a central issue: 'US consultants accused in row over fat cat pay' headlined an article in the London *Evening Stan-dard*: 'A firm of shadowy US consultants has emerged deep at the heart of the row over British boardroom fat-cattery' (Armitage, 2003), and a general secretary of the trades union Amicus, Roger Lyons, spoke of 'These enormous American-style remuneration packages . . . This has to be wrong' (cited in ibid.). Perhaps the question of class, though along new lines, might find its way back on to the political agenda. If it does so, however, we shall need to be far more aware than has been the case in the past of the complex geographies in which it is embroiled and that both underpin and flow from it.

It may be beyond passé – but we'll have to do something about the rich

If you want to be deeply unfashionable, just read on. If you want to enter terrain so wildly out of date that mere mention of it has become taboo, then you've come to the right place. Brace yourself. Late last month two bankers strode into Umbaba, one of London's most modish watering holes, and asked the bartender to fix them a drink. Not any drink, you understand, but the most expensive cocktail he could concoct. He set to work, blending a Richard Hennessy cognac that sells at £3,000 a bottle, Dom Perignon champagne, fresh lemongrass and lychees – all topped off with an extract of yohimbe bark, a West African import said to possess aphrodisiac powers. He called it the Magie Noir – and he charged £333 a glass. The bankers ordered two rounds for their table of eight. Their final bill for the night: £15,000.

Those same men, or their colleagues, may well have invested £200,000 in a Bentley or Aston Martin, or they might have paid celebrity hairdresser Nicky Clarke £500 for what the salon describes as an "aspirational haircut". They are the customers sought by the London estate agent who offers a three-bedroom flat in Kensington as a "starter home" for £2.25m. They are the target reader of the newly launched Trader magazine, with its ads for private jets or five-storey yachts (complete with submarine).

This is the world of the super-rich, financiers pulling in salaries and bonuses in the millions, and sometimes tens of millions, of dollars. They are partners in hedge funds and private-equity firms – buying, selling and gambling in jobs that most mortals barely comprehend. They spend money on vast estates or wild fancies. Sometimes the splashing out is literal: a favourite pastime is spraying champagne in the manner of a formula one winner. (In August one London banker fizzed away £41,000.)

Nothing new in all this, you might say. The rich, like the poor, are always with us. But that would be wrong. Robert Peston, City editor of the Sunday Telegraph, estimates that this year no more than 200 to 300 hedge-fund managers will carve up $4.2bn of pure profit between them.

Extract from article by Jonathan
Freedland, *The Guardian*,
23 November 2005

Capital delight indeed

2

'A SUCCESSFUL

CITY, BUT . . . '

This exuberant, champagne-swilling claim of the success of London's reinvention is, however, almost always hedged about with a regretful caveat – 'but there is "still" poverty too'. The language of this caveat is in itself revealing. Or, rather, it is systematically *un*revealing. It is a language of reservations, of howevers, and of paradoxes. Even the excellent *London divided* begins this way: 'London is the most dynamic, cosmopolitan and diverse of our major cities, and one of a handful of truly "world cities". . . . *However* London's formidable wealth generating capacity coexists with truly staggering levels of economic disadvantage' (GLA, 2002, p. ix, emphasis added). The London Chamber of Commerce and Industry deploys both 'however' and the notion of paradox: 'One of the world's leading centres for global business, it is *however* a capital with a vast disparity of wealth. *Paradoxically*, whilst full-time earnings in London were 37% above the UK average for men and 31% for women in 2002 . . . and it has the highest gross disposable household income per head, London suffers from high levels of deprivation' (Hill, 2003, p. 5, emphasis added). Hamnett, documenting 'the unequal city', argues that London 'exhibits the paradox of great wealth and considerable poverty in close juxtaposition' (2003, p. 19). And Ackroyd, too, remarks upon this 'paradox' and treats it, rather vaguely, as representing the contradictions of the human condition (2000, p. 766). My dictionary (*Webster's seventh new collegiate*) gives, as the most relevant definition of paradox: 'a statement that is seemingly contradictory or opposed to common sense and yet is perhaps true'. It would seem from the citations above that current common sense entails a belief that growth of any kind, eventually, is good for all; that because some are rich it is surprising that others, who live close by, are poor. Likewise, the juxtaposed coexistence of vast wealth and deep deprivation is imaginatively constructed, through the distancing terminology of 'however', as though the two were separate facts.

In reality these two aspects of London – the success and the poverty – are intimately related. First, they are the combined outcome of the politico-economic strategy of neoliberalisation – the establishment at national level of what Hall has called '[t]he two-tier society, corporate greed and the privatisation of need' (2003, p. 12). Second, the establishment at national level of this society has found its most acute expression in the capital. It is here that greed and need most obviously encounter each other. And third, the fact of this concentration in London (in other words the very geography of inequality) has produced, through the dynamics of the negotiation of place, its own, further, effects.

Some facts are indisputable. Inequality between rich and poor, the glaring starkness of class difference, is more marked in London than anywhere else in the country. Unemployment in Inner London is higher than in any other subregion in England, while Outer London hovers around the national average; on almost any index you like to mention there is an enormous geographical variation between boroughs (GLA, 2002, p. xii).[1] London has the highest incidence of child poverty, after housing costs have been taken into account, of any region in Great Britain (xi). The gender pay gap is wider in London than in Great Britain (xiii). London has local authority areas with both the highest and the lowest rates of means-tested benefit receipt in the country (xiii). 'Nearly a quarter of London's children (24 per cent) are living in households dependent on Income Support' (xiii) (the rate for Great Britain as whole is 16 per cent, and London's rate is the highest of any region). Among pensioners, too, poverty is common – in Inner London, a quarter of people aged sixty and over are on Income Support (15 per cent in Outer London and in Great Britain) (xiii). Homelessness and overcrowding are higher in London than elsewhere. And such pressures feed through in a host of ways to daily life – in stress, in isolation, and in the endless grinding difficulty of making ends meet (Buck et al., 2002, especially chapter 7). These are indicators both of poverty and of inequality. All of them are patterned and cut through by space, gender and ethnicity. On the first of these dimensions, for instance, the difference in that most basic of indicators, life expectancy, is stark even between the boroughs of London. On average, women in Kensington and Chelsea live nearly six years longer than women in Newham; and men in Kensington and Chelsea (again) live nearly six years longer than men in Southwark (xv).

Such inequalities are not new, but they are being reworked and in some ways (in the midst of London's success) becoming more marked. Hamnett pulls out four axes of recent changes:

First, that the incomes of high-income groups have risen much faster than those of the poor, and that inequality has consequently increased dramatically in London since 1979. *Second*, that the relative size of the rich group has also increased substantially, . . . *third* . . . that the greatest increases in earnings have been amongst those working in the City of London, which points to the key role of financial and business services in increasing inequality. . . . *fourth* . . . that the inequality in household incomes has risen substantially, and that both the incomes and the share of total household income of the lower groups have declined substantially while those of the richer households have increased. (2003, p. 77, emphases added)

This is not the violent polarisation or the fierce systematic exclusion that blights cities in the USA. There has been, over the years, considerable argument over what has been dubbed 'the global cities hypothesis', which postulated a growing polarisation into an hour-glass income distribution and social structure in such cities.[2] London does not simply conform to this hypothesis – the specificities of place are anyway too marked to allow the simple reproduction everywhere of such a generalised outcome (Amin and Graham, 1997). As Hamnett (2003) points out, while the gap between rich and poor has grown, and the numbers of the former have expanded, there has not been such a growth in numbers of the poor; and while earnings have risen fastest for already high-earning groups, they have risen too for all groups. Similarly, Buck et al., in their minutely detailed investigations, 'did not find a major growth of the "truly excluded" in London'. One reason for this 'is the relatively wide "safety net", including the stock of social housing' (2002, p. 257). As they argue, 'in comparison with some Western European welfare systems, Britain's has done less to combat deprivation

and exclusion', but 'London is unlike Chicago: here, neither
housing nor job markets are so divided (primarily by "race"),
nor is the welfare state so reduced or family and social networks so broken, as to extend this limited area-based deprivation and exclusion across large parts of the inner city – or
even to suggest that this process is underway' (ibid.). This is
not, then, some inexorable working-out of an undifferentiated neoliberal capitalism at the level of the city-region, as a
generalised 'global cities' thesis has on occasions postulated.
It is a unique articulation: a place where market capitalism is
in part produced and propagated, yet where it is also still
embedded in (the remains of) a social democratic settlement.
Politically, and this is why the argument is important, this
points to the need to defend the welfare state. It also points
once again to the double spatial positioning of London in
relation to influences and characteristics of both the USA
and continental Western Europe. There has, nonetheless,
been a notable increase in inequality. There are, moreover,
some intimations of polarisation as such. The most recent
data indicate 'a small but significant growth in the absolute
number of jobs at the bottom end of the occupational hierarchy' (May et al., 2006, p. 6). Such jobs are largely filled by
new migrants, from the global South and also increasingly
from the Eastern European countries newly acceded to the
European Union. For while with one arm the British state
continues to lend, through the welfare system, some protection to some (even while undermining over three
decades the ability of trades unions to provide representation), with another arm, through the expansion of the EU
into a simple free market, it has increased pressure on the
lower paid (Flynn, 2005). The London Chamber of Commerce and Industry, in its *Policy guide* for 2004 (the year of
accession), welcomed increased immigration to London
'particularly from the new EU accession countries' (2004,

p. 13). Again, London is entangled in the complex geographies of neoliberalism. As Clark points out, what the people of the old communist bloc are fleeing are the neoliberal polices in those countries, where GDP has fallen and where both inequality and unemployment have risen spectacularly. As he says: 'Western politicians laud the countries of "new" Europe for their "dynamic, flat-rate tax" economies, but deny there is any link between the economic reforms and the massive exodus' (Clark, 2006). Moreover, as a legacy of the communist era, many of these workers are highly skilled and educated. The biggest losers are the East European countries themselves, but the movement puts pressure also on low-paid workers in London. As Mervyn King, governor of the Bank of England, reflected: 'Without this influx to fill the skills gap in a tight labour market it is likely that earnings would have risen at a faster rate' (quoted in Elliott and Moore, 2005). Thus are inequalities within the global city again intimately related to the inequalities in the world beyond.

Yet, even this emerging intimation of polarisation has its specific contours. 'Within the global cities literature the demand for migrant workers tends to be attributed to either a growing demand for low-wage workers to service the high-income lifestyles of an expanding cohort of professional and managerial workers (Sassen, 1991) or to an expansion of informal economic activities and of the "grey economy" (Cox and Watt, 2002; Samers, 2002). Both may be important. But in London at least the demand for such workers goes much deeper' (May et al., 2006, p. 22). Most of these workers, in London, are in the formal economy, including in the public sector.

One question all this immediately poses is what it *means* for a city to be successful. Can a city with such inequalities within it be deemed to have succeeded? How is success to be measured? The measure most frequently deployed contains, even within its seeming statistical objectivity, a real class irony. This is productivity per head. On this measure, there is no doubt that London stands out, both in the UK and more widely in Europe. But, quite apart from the inevitable problems of measuring 'product' in manufacturing industries, in services these problems are compounded. One bizarre result is that the high salaries in London contribute to raising this measure of 'productivity'; and increases in salary (the directors' 16 per cent) raise this measure of London 'success' over and above that of other UK regions even more. The more they earn the higher their productivity. We should, then, treat this measure of success at least with a degree of irony. Beyond this, however, 'product', even were it to be properly measurable, may not anyway be an adequate index. International surveys of liveability rarely find any of the major world cities to be among the most highly rated – more likely it will be the likes of Barcelona, Sydney and Stockholm that shine. Indeed, the mayor has consistently tried to recognise this in relation to London. The very notion of negotiation of place in the midst of such cultural diversity was clearly what came to the fore in the aftermath of the bombing, and clearly too this is what, in opinion polls, is rated London's best asset amongst its inhabitants. From the beginning of his tenure Ken Livingstone has stated his aim as being moving towards an exemplary sustainable world city. Moreover, those measures of productivity per head not only include products both noxious and benign but they also squeeze out consideration of non-market activities, including unpaid domestic labour – the London Industrial Strategy of 1985 estimated that this was 'the most important sector in the economy in terms of labour time expended' (GLC, 1985, p. 19). And of course they say nothing about distributional criteria, and, in a city as unequal on so many measures as is London, distribution must surely be not a caveat to an achieved success, but part of the criteria of success itself. And finally, more tentatively, and looking ahead a little, there is the new emphasis on, and the beginnings of a calculus of, happiness and well-being (see, especially, Layard, 2005; and Purdy, 2005). As Purdy argues, this still raises questions of distribution, and of the significance of other criteria such as personal autonomy and social justice, but 'if happiness cannot simply be substituted for GDP as the lodestar of public policy, its claims can hardly be disregarded' (2005, p. 142). This is particularly so in an era when the triumph of market individualism is increasingly understood to have resulted in a 'social recession' (see, for instance, the initiative on 'the good society' published by Compass, 2006). Local Futures Group's (2006) *State of the nation* report, documenting the geography of well-being in Britain, shows how a shift towards such criteria at least begins to trouble the accepted simple geographies of 'success' (see also Soper, 2006; Commission on Urban Life and Faith, 2006). The point is simply that there is a question: what, anyway, does it signify to dub a city, or a city-region, successful? What is a city, or a city economy, *for*?

And there is a further issue, which is that the history of growth along these lines, since London's reinvention, has shown no signs of alleviating the inequality within it. In part this must be due to the very attractiveness of London, in combination with poverty elsewhere, and the consequent immigration from other countries – not only their arrival in itself but the depressive effect on wages at the lower end. But, as we shall see, it is not only this; rather the persistence of such inequality in the midst of 'success' must put a question mark over strategies that call – in the name of reducing poverty – simply for more growth of this same kind. 'Growth' can take different forms, with differing distributional outcomes.

In fact, the very terms of London's reinvention – the nature of its particular form of growth – are part of the dynamic behind this reproduction of inequality. There is a tension at the heart of the London economy, between this (kind of) global-city status and other parts of the capital's economy and society. It is not the case that 'London is a successful city, but . . . (there are still problems of inequality)'. Rather it is that London is a 'successful' city, and *in part as a result of the terms of that particular form of success*, inequality is reproduced within it.

One obvious way in which this occurs is simply through the structural fact that some aspects of this global-city growth place constraints on, and present obstacles to, the growth – sometimes even the survival – of other parts of the economy. The usual view of London's changing economy would have the decline of manufacturing and dockwork happening alongside (or sometimes being compensated for by) the growth of employment in finance and business services. It is as though they were independent phenomena. Of course, they are indeed both part of wider long-term trends, but in the context of the material juxtapositions of place they are by no means independent. This is particularly evident in the case of land and property prices, which have rendered unsustainable a whole range of jobs in manufacturing and interstitial services. The pre-emptive land purchases around the fringe of the financial City are an obvious case, but the effects are more general. The London Voluntary Service Council, for instance, in their response to early versions of the *London plan*, suggested that 'rising property prices have had a detrimental effect on voluntary and community organisations, with many priced out of their local area' (London Voluntary Service Council, 2002). And the outward spread of the old City is precisely into some of the most mixed-use areas of the capital. The effect is to denude London of some of that very diversity that Buck et al. (2002) have pointed to as the capital's greatest asset. Many of the cities of the North (Sheffield, for instance) have long been lambasted by comfortable commentators for overdependence on a single economic sector. Checkland (1976) wrote evocatively of the effect of the upas tree, beneath whose spreading branches nothing else can grow (he was writing of Glasgow). The finance constellation might turn out to be, though working through very different mechanisms, the upas tree of tomorrow's London.

Moreover these terms of London's reinvention have other effects too, in contributing further to the decline in jobs in London for manual workers. Since 1992, virtually all the increase in full-time employment in London has been in occupational groups that require a university degree or equivalent. On the one hand this means that many of these posts are filled by graduates migrating into London (many of them coming from the North and West of the UK and thus exacerbating the difficulties those regions have in kick-starting growth – see Part II), and thus contributing to the population growth of London and the South-East. On the other hand it means that resident Londoners without such qualifications are simply bypassed as 'not relevant' to the central dynamic of the capital's contemporary economy. This tension at the heart of London's current form of growth is recognised in the mayor's consultation paper *Tackling poverty in London*. In its thoughtful analysis of the pattern of disadvantage in London, this paper identifies a set of factors specific to the city that are responsible for the continuing inequality. The second of these (out of five) is:

the pattern of demand for labour:
London's labour market has changed dramatically over the last twenty years, with massive job losses in manufacturing and massive growth in service sector employment,

particularly in higher skilled jobs. Virtually all the growth in full-time employment in London in the 1990s was in managerial, professional and associate professional occupations. These jobs generally demand third level qualifications. The increased concentration of employment opportunities in higher skilled (and higher paid) occupations risks leaving many Londoners with low or no qualifications excluded. (GLA, 2003, p. 3)

The reinvention of London along these particular lines, in other words, contributes to unemployment for some even while it expands opportunities for others.

The document also recognises the pincer movement of migration in which London's poor, and those without recognised skills and qualifications, are caught. It identifies the third regional factor which is important in explaining the pattern of disadvantage in London as:

> *the openness of London's economy*:
> The growth of highly skilled employment partly reflects the ability of London's labour market to draw on skills throughout the UK and internationally. On average over 150,000 people move into London from elsewhere in the UK every year, while London also receives more than half of all international immigration into the UK. This means that London residents face an extremely competitive labour market, with higher risks of long term exclusion for those who are disadvantaged. (GLA, 2003, pp. 3–4)

London's poor, in other words, and those without higher-level skills, are caught in the cross-fire of the city's reinvention. On the one hand the employment generated by the new growth is not for them; it is a different class project entirely. This is London 'global city' as a capital of neoliberalisation. On the other hand, as their old jobs disappear, other workers arrive from around the planet to compete for the few economic opportunities that do remain for them. This is London 'global city' as cultural mixity. The old London working class (already ethnically mixed), caught between the two world citydoms, feels itself under threat from both directions.

Moreover, these geographies of migration within which London is set operate differentially. While those working in jobs at the lower end of the wages spectrum find themselves under competitive pressure, as the workings of the market both push down wages for some and threaten exclusion for others, the irony is that the in-migration at the very top of the salary spectrum does not have this effect. As Lansley writes: 'does such a global market [for corporate bosses] really exist? For the most part, the explosion in executive salaries has been a largely Anglo-Saxon phenomenon' (2006, p. 94) . . . 'As one observer has put it, "it is not the invisible hand of the market that leads to these monumental executive incomes, it's the invisible handshake in the boardroom"' (ibid., p. 95).[3] The propounders of the philosophy of market competition, in other words, do not themselves submit to it. Those who would argue the market-rationality of top salaries resort to one of three hypotheses: the globalisation hypothesis, the hypothesis of skill-biased technological change, and the superstar hypothesis. According to Krugman (2002), although each may make its contribution, none is adequate. Rather what is at issue is the role of social norms in setting limits to inequality: and old norms that may have frowned upon deep inequalities and explicit expressions of individual greed have been eclipsed by what he calls an ethos of 'anything goes'. Once again what is at issue is the tectonic shift between forms of social settlement.

It is these highest salaries that make London the most unequal city, and London and the South-East the most unequal region, in the UK. The bottom decile of wages is far more similar across the country than is the top, in other words, while the highest paid in London/the South-East earn far more than the highest paid in other regions. There is also a clear gender difference. For men, the inequality in hourly earnings between top and bottom deciles is far greater in London than nationally. The earnings distribution for women in London is less unequal, and much more like the national distribution. Moreover, as we have already seen, since the 1990s lower-paid workers have had much lower growth in real earnings than have those in higher-paid groups (GLA, 2002). The group statistically responsible for the peculiarly acute inequality in London and the South-East, in other words, is highly paid men.

This is the case nationally as well. However, the geography of this group within the nation matters deeply. *First*, the overwhelming concentration of the very wealthy into London/the South-East sharpens the inequality both between North and South within the country and within the city and region. *Second*, their geographical concentration into a self-referential echo chamber reinforces their distance from the rest of us. And *third*, that spatiality (the 'paradox' of the juxtaposition of rich and poor within the city) has effects. The expanding presence and prosperity of this new elite feeds out to all Londoners. Most obviously, there is the effect on housing costs. Hutton concentrates on the effects of these extremes of inequality at the upper end: 'The upsurge in house prices, for example, in the smart parts of British cities and especially in London, effectively pricing them out of the reach of most income groups, is closely related to the new patterns of American-style executive pay transmitted by winner-takes-all effects throughout the senior managerial job market. The impact of income inequality on the American model in a country whose land area is very much smaller creates extraordinary pressure in the land and housing markets, with disastrous social results' (2002, p. 38). Buck et al. refer to professionalisation more broadly: 'The high demand for professional, managerial and related workers in London drives up housing costs generally and supply constraints increase this effect' (2002, p. 243). Both the very high salaries and the expansion of the group, in other words, that is *both* the 'sectoral' nature of economic change within London (the shift towards finance and business services) *and* the social nature of that change (what Hutton calls the American model) are at fault. Moreover, the 'supply constraints' that Buck and his co-authors are referring to go beyond the problem of restricted land supply as adumbrated by Hutton (and anyway itself exacerbated by the practices of the property and building industry), to include also the sale of council houses – another element of the deregulation/privatisation economy upon which 'London' has thrived. 'Low-income households are then faced with either paying high prices, even for poor properties, or trying to gain access to that part of the housing system which is relatively sheltered from these effects [rising prices]. However, in the past 20 years this protection has been weakened by social housing sales which have also led to the marginalization of much of the remaining council housing – frequently this is now some of the worst housing' (ibid.). It is not just, in other words, that the poor in London have to live ('paradoxically') cheek by jowl with the rich; it is also that that very co-presence makes their lives even harder. Juxtaposition in place makes a difference. Nor is it only absolute levels of poverty that are at issue; the fact of inequality in itself, here within the context of the negotiation of place, has its own autonomous effects. As Jackson and Segal write: 'In the developed world, inequality is more important than per capita GDP in determining the living standards of the poor' (2004, p. 5; Wilkinson, 2005).

However, again the geographies of inequality affect the workings of this relation between poverty and inequality. Here within London it is possible to see some of the mechanisms produced by the juxtaposition of rich and poor within an individual place – indeed in one of the most 'successful' places in the world economy.

As *London divided* puts it, 'taking account of housing costs radically alters the distribution of household income. *This impact is particularly marked at the lower end of the income distribution and in London*' (GLA, 2002, p. 12, emphasis added). In London, and especially in Inner London, the difference in levels of income poverty between before (BHC) and after (AHC) housing costs tends to be greater than anywhere else. Were all the indices of poverty mentioned at the start of this chapter to be presented on an AHC basis they would be even worse. (And this, of course, is the measure that London groups and representatives tend to use when bargaining with national government or with other regions over their share of the national cake – see Part II.) Housing costs are higher, in other words, in comparison with those in other regions than are wages for the lower paid. Moreover this pressure of housing costs feeds through to housing conditions, to levels of homelessness and overcrowding, and to the stresses and strains that accompany these things.

There are other ways, too, in which through the dynamics of place the high incomes for some, and the fact of inequality in itself, create extra difficulties for less well-off Londoners. In part, such difficulties are also the consequence of the sheer growth of London. The cost of transport is one evident example. So too is the cost of childcare. Historically, women in London have been more likely than women elsewhere in the country to be economically active. Recently, however, this historical pattern has been reversed. As women elsewhere in the country have joined the active paid labour force in increasing numbers, the levels in London have stayed static. Most analyses concur that 'women with children face particular difficulties in London's labour market due to high housing and childcare costs' (GLA, 2002, p. 7). The GLA household survey shows that lack of affordable or suitable childcare is cited as the reason for not working by 31 per cent of women who express a wish to work (ibid., p. xii). Or, again, the particular combination of levels and distributions of wages and cost of living with a welfare benefits system that is set nationally produces a peculiarly London version of the benefit trap. This is a specifically spatial trap – once more, place matters. On the one hand, London's relatively high wage rates and high costs reduce the eligibility there [for benefits] of the working poor (what Buck et al. refer to as 'a telling example of a negative relationship between economic dynamism and social cohesion'; 2002, p. 257). On the other hand, because of the high cost of living, and the cost of getting to work, people currently dependent on benefit can find gains from entering employment completely eroded by the loss of entitlement to benefits, particularly housing benefits (GLA, 2001, p. 12). Such mechanisms further exacerbate the problem of unemployment, itself a crucial component in the changing patterns of inequality (see Buck et al., 2002, p. 156). Put together, this higher cost of housing, transport and childcare is ranked by the GLA study as the first among the specifically London factors which are important in explaining the pattern of disadvantage in the city (GLA, 2003, p. 3). What it is less explicit about, however, are the *causes* of this higher cost of living. It is not just 'a regional factor'; it is in part a product of the very nature of growth *in* the region. The very nature of the 'reinvention' of London and the growth strategy currently being pursued themselves contribute to the poverty and inequality in the city and its region. As Edwards has put it,

London has a dual character, 'as simultaneously a wealth machine and a poverty machine, as sets of relationships which produce exploitation and deprivation as surely as they produce growing net outputs' (2002, p. 29).

One response to this increase in inequality has been to attempt to divide it up into that attributable to national effects and that attributable to the region itself. It is an operation performed statistically, through normalisation and comparison of London with the rest of the country. Buck et al., for instance, conclude from this kind of analysis that both effects are in operation, but with the national effects playing the major role. The GLA, coming at the issue from a purely policy point of view, and after having pulled out the five regional effects already referred to, nonetheless concludes that 'The powers and resources available to regional and local government to address these issues are limited. . . . The main role in reducing poverty rests with central government' (GLA, 2003, p. 5).[4]

One problem with purely statistical approaches to these issues, and with all approaches that so sharply distinguish between 'scales', is that it is difficult to acknowledge mechanisms that connect the categories. For in fact these scales are intimately intermeshed. 'National' changes may show up in exacerbated form in London, for instance. Since the 1980s the national governments' lack of commitment to equality has produced more marked effects in London and the South-East than elsewhere. On the other hand, some of the dominant pressures, and voices, *producing* the shift in national economic policy of which this lack of commitment to equality was a part, emanated precisely from London and from the specifics of the London economy and class structure. National and London effects are in this sense quite entangled. What is not in doubt, however, is that there are mechanisms internal to the working of the economy of

London and the South-East – the effects of the material proximities of place – that have contributed to the growing poverty and inequality. It is, moreover, both poverty *and* inequality that are at issue. Tony Blair's conviction that, 'while poverty matters, wealth does not' (Lansley, 2006, p. 196) is wrong even as a general proposition. In the context of a place such as a city it is doubly so. The business of 'tackling poverty' by no means requires a focus only on the poor. 'Greed' and 'need' are not independent of each other. The presence of the very rich and the sharp shift in the economy towards professionalisation of this over-remunerated kind are integral to the mechanisms of poverty production. And in that there are indeed strategic questions to be posed about the balance and direction of the growth of London itself.

Finally, this is a kind of growth that has made it extremely difficult for London to reproduce itself.[5] Low-paid workers in both public and private sectors find it difficult to survive, and are deterred from taking jobs by the combination of the high associated costs and the benefits trap. Report after report documents the difficulty of recruiting public-sector workers to the police, to the health service, to education. Here too national policies as well as local economic strategy are at issue, in particular among the former the low level of the minimum wage and the holding down of public-sector salaries. As a result a whole set of non-market interventions has to be introduced, simply to keep the city, and its region, running.[6]

There are, then, tensions at the heart of this reinvention of London, mechanisms that result from this model of growth and that operate through within-place dynamics. They make more difficult the daily reproduction of the place; they exacerbate the problems of non-dominant parts of the economy and society; they sharpen the poverty and the inequality. London-world-city as global centre of articulation of a neoliberal world economy poses huge questions of 'redistribution' that are utterly entangled with those of 'recognition' that dominate the characterisation of London-world-city as multicultural (see Hall, 2000). Not only are the inequalities themselves sharply cross-cut along lines of ethnicity, but the dynamics of London as neoliberal global city and the politics that have accompanied it at national level could undermine the relative mutual toleration, the 'convivial culture' of mixity (Gilroy, 2004) that was so widely celebrated in the aftermath of the bombing. The two different narratives of London-world-city may in other words come into contradiction. In 2006 the far-right British National Party gained seats on local councils in some working-class areas of east London. It has to be read as a protest, exploiting a political void, a point of tension to which no other party is responding, by those caught in the cross-fire of the double movements of globalisation, whose voices and interests are systematically unattended to by many of the promoters of the neoliberal project (Weir, 2006). Some of the problems of poverty of which so many of the representatives of London now complain are in this sense and in part problems both of national economic strategy and of its own making. They are integral to the new social settlement. The implied logic behind 'London is successful, but . . . ' conceals the fact that in many ways the problems are an integral part of the functioning of London in its new, reinvented, form.

3

IMAGINING THE CITY

Places are crucial. Huge shifts in politics, culture, economics, do not happen evenly. This was particularly true with
the establishment of the current neoliberal consensus. It is
often pointed out that in the USA the battle over New York,
and the huge fiscal crisis which that city underwent, was a
forging house, and laboratory, for neoliberal practices that
would eventually be generalised (see Tabb, 1982; Zevin,
1977; Harvey, 2005). So too in Britain, though in a very different way, what happened in London was central to the
national transformation.

The emergence of this new London must be understood
in its longer historical perspective. In part this is because it
enables the uncertainty, and the unevenness, of the change
to be appreciated. But it is also because this very appreciation underscores the fact that the process was not inevitable.
The triumph of neoliberalism was struggled over, both
nationally and in particular places. As Rustin writes, 'The
effect of the forgetting of history and the opposed traditions,

is to "normalise" the present as the one imaginable world of all possible worlds' (2006, p. 4). It is politically debilitating.

In the United Kingdom, from the mid-1960s when national employment in manufacturing reached its peak and began what was to be its inexorable decline, through – at an international level – the collapse of the Bretton Woods agreement and the battle over oil prices in the early 1970s, the story is that of the breakdown of the postwar social democratic settlement in the UK and the various attempts at, and struggles over, finding an alternative. As Devine sets out the options: 'In this historical conjuncture, two alternative post-social democracy trajectories presented themselves: a move in the direction of economic democracy, building on the gains of the long boom, as a transitional stage towards socialism; or a move to neoliberalism, reversing the post-1945 gains' (2006, p. 152). These were certainly the alternative options at either end of the political spectrum, but in practice the ensuing decades were a period of trial and error and of complex political struggle. The general story is well known (see, for instance, Gamble, 1981).

What is less well appreciated in this oft-told story is, in general terms, the significance of its geographical unevenness, and more particularly and concretely the fact that the current victory of neoliberalism both represents a victory of (a part of) London/the South-East over the rest of the country and, for that very reason, means that the conflict over London, and what kind of a city it was to be, was (and remains) crucial to the national outcome.

The breakdown of the old class settlement and the conflictual establishment of the new was always an uneven one – economically, culturally, ideologically and politically – sharply differentiated between regions, between major cities and the rest of the country, between London and the South-East, and between London-and-the-South-East and elsewhere. And it was historically hesitant. But in each period (the Wilsonian modernisation of the 1960s to 1970s, the muddling through and defeat of the old in the 1970s and 1980s, and the emerging hegemony of neoliberalism) geography and uneven development played a distinct role. And each period contributed – though in different ways – to perpetuating and reworking forms of geographical inequality.[1]

The 1960s and early 1970s had seen a last attempt to hold things together, under Wilson and Heath, through a combination of high-tech modernisation, the running down and reorganisation of heavy basic sectors (shipbuilding, coal, steel), and a significant expansion of the welfare state. This was an attempt at an alliance between the state, the Trades Union Congress (TUC), big manufacturing capital and new high-technology sectors, and significant proportions of the growing white-collar strata (Massey, [1984] 1995). The contrast with today could hardly be more marked. In that period, not only was a very active regional policy seen, in the National Plan no less, as an explicit arm of the national accumulation strategy (largely by expanding the labour force), but capital itself, in parts, used geographical mobility as a means of addressing the crisis that was increasingly overwhelming it (both by moving abroad – this was the early period of the establishment of 'the new international division of labour' – and by moving within the UK, both from South to North and, increasingly, out of the inner cities).[2] The combination of this with increasing employment in the public sector (the only significant part of the economy to be relatively evenly spread, in terms of numbers of jobs, across the country) led to a closing of the North–South divide in terms of the classic measures, in particular unemployment. But the modernisation project also involved an increasing centralisation of control over accumulation and a significant expansion of 'high-technology' sectors and employment.

Together, these developments meant that, while the old regional problem (defined in terms of unemployment) became temporarily more muted, a different form of inequality between North and South – that between control and execution – was intensified (Massey, 1979). A remoulded national spatial division of labour, a geography of inequality based on the spatial reorganisation of an anyway changing social structure, was being established. From Bristol to Cambridge, the 'sunbelt' of newer, higher-technology, industries was on the rise. Bolstered by previous state investment in defence establishments, this stretch of country was also favoured by expanded technical and professional classes that knew, and wielded, their locational power. This was a geography moulded by class. Given the environmental preferences of these expanding strata, new growth was not going to take place where it was needed, where there was decline. Decline indeed, and the presence of a blue-collar working class, was a serious locational deterrent. It was a locational preference exerted, too, by 'top' civil servants who resisted all attempts (as they still do) to get them to move away from their favoured London and the South-East. By the 1980s, that aspect of the national divide was well established (Massey, 1979, 1983, [1984] 1995). In 1985 *The London industrial strategy* of the then Greater London Council put it thus: 'London is . . . the head which directs the hand of labour on a national and international scale' (GLC, 1985, p. 334).[3] And Dorling and Thomas's (2004) mapping of the 2001 Census, and comparison with that of 1991, demonstrates that the process has continued – there is further growth and concentration towards London of both corporate managers and professionals (see their chapter 6). Moreover, and as we shall see providing even further reinforcement, unlike in the case of New York as global city (which anyway does not so dominate the national distribution of headquarters and financial institutions), this constellation of private economic power in London sits right next to Westminster, seat of national government. The upper echelons of all this form an increasingly prominent apex of a grotesquely unequal national grammar of power and prosperity (Amin, Massey and Thrift, 2003).

Moreover, Peck and Tickell (2002), in their analysis of the establishment of neoliberalism, distinguish between an earlier period of roll-back (which might be taken to include in part the collapse, as well as the deliberate destruction, of the old class compromise) and the beginnings of the construction (roll-out) of a new regime: the old had to be destroyed before the new could be installed. The 1970s, then, witnessed the collapse of the old settlement, while the 1980s, as we have already seen, was the decade in which, in the iconic moment of the crumbling of the miners' strike but also more widely, the old historic bloc of labour was defeated (Devine, 2006). It was in the later 1980s and in the 1990s that the direction of movement shifted, as the dismantling of the terms of the old settlement began to be accompanied by the consolidation of a new hegemonic set of forces. This new balance of forces – the elements of the establishment of a neoliberal settlement – began to be consolidated in the 1990s under the Tories and the project has been continued under New Labour. Indeed, Devine argues that, whereas 'Thatcherism had destroyed the old historic bloc and created the basis for a new neoliberal era, it had not yet succeeded in creating a new historic bloc in which neoliberal principles and policies became the generally accepted ideological cement holding it together. This was to become the historic mission of New Labour' (ibid., p. 158; see also Hall, 2003).[4] Each of these movements had its geography. The roll-back swept away much of the mining and manufacturing of the North, the cities, the mining regions and the old East End. The assertion of the new, in terms of economic

and employment growth, was overwhelmingly concentrated
in London/the South-East, reinforcing the new spatial division of labour that had begun to emerge from the Wilson
period on. When Thatcherism arrived, at the end of the
1970s, it by no means swept the board, even among capital
(Massey, [1984] 1995). The base of Thatcherism was overwhelmingly in the South-East. This was so in two ways. First
it was from here that emanated the impetus for the new
neoliberal settlement. Second it was here that cultural and
political consent was most easily achieved. Nor was this
simply a reflection of the changing social composition
(though this was part of it, as the expanding private-sector
professional strata increasingly congregated in that corner of
the country). It was also that among both skilled workers and
professional and managerial strata there was regional political differentiation – in both cases location in the South-East
tending to be associated with a more pro-Thatcher position
(Price, 1979; Penniman, 1981). In other words, geographical
differentiation was itself a significant element in the nationally registered process of class de-alignment. Moreover, the
serious *challenges* to Thatcherism also had clear geographical
bases. On the one hand there were the defensive battalions
based in the old industrial, and particularly the mining,
regions. On the other hand, there was a new kind of more
creative resistance in the cities, as the new urban left tried to
imagine a way out of the impasse. The 1981 local elections
had sharpened the North–South political divide (at the same
time as reconstituting it, as the West Midlands – previously
more politically ambivalent – found itself in the throes of
deindustrialisation, not as North–South at all but rather as
London-and-the-South-East versus the Rest), and within this
shifting overall pattern the cities were highlighted as points
of Labour red within the wider ocean of Conservative blue. It
was in the cities – in battles over funding, over rate-capping,

and over alternative economic and social strategies – that new bases of opposition were established (see also Toulouse, 1992). The inter-place twinning solidarities of the miners' strike, and the meeting of Arthur Scargill and Ken Livingstone, seemed at the time potentially momentous – the beginnings of conversations between the old resistance and the tentative experiments with the new. Within months, both wings had been defeated.

Geographical differentiation, then, was more than some result of the spatial distribution of national phenomena. It was an active element in the troubled production *of* those national changes. Moreover geographical imaginaries were evoked to feed the politics. The wider South and East, excluding most of London, was held out as the paradigm (individualistic, 'enterprising', suburban, owner-occupying, car-owning) of what the rest of the country might become if only it could change its ways (Allen, Massey and Cochrane, 1998; Hudson, 2006b). Thatcher was famously pictured trudging a wasteland in the North-East, was recorded calling on Northerners and Scots to become more entrepreneurial, and was heard determining to 'take back the cities'. In this context, New Labour's 'historic mission', especially given its electoral bases in both cities and regions, can certainly be seen as creating a new hegemony (Devine, 2006), but this was in part precisely by broadening the *geography* of the acceptance of, or at least reconciliation to, the terms of the neoliberal settlement – in other words, the task was importantly also a specifically spatial consolidation. It might additionally be argued that it has achieved this (insofar as it has done so) very much *because* of this old regional pattern of attachment, as well as because of the lack of alternatives at national level, for, as Part II will go on to argue, New Labour's own real loyalties have lain – in keeping with its neoliberal proclivities – with London-and-the-South-East and with what, in social, cultural and ideological terms, that part of the country represents (see Allen, Massey and Cochrane, 1998). This way out of the impasse, then, was a victory of London/the South-East. But it was to have effects which reverberated through the country as a whole, including in particular the further dismal perpetuation, and continuing reinvention, of its regional inequality.

In that sense, the conflict over what was to happen *in* London, in the 1980s, was critical. In part this was simply because it encapsulated in heightened form the wider picture – it was iconic. But it was also critical because, had the political outcome in London been different (had the city's reinvention been on other terms), so too the national outcome would not have been as it is today. London is what it is today as the result of the victory of neoliberal, deregulatory forces dominated by a City that had long favoured such a perspective and that now, with the Keynesian, social democratic and relatively egalitarian settlement so plainly in trouble, saw its chance and took it. The forces against it were weak and divided. While neoliberalism was wholeheartedly supported by its 'natural' political party, Labour was riven by dissent, its dominant voices hostile to any radical democratic way out. The national-level Alternative Economic Strategy (AES) of the 1970s was destroyed in this conflict. However the AES, so often credited with being the last attempt to propose a left alternative, was in fact followed by the even more challenging new urban left of the 1980s, with their place-based strategies for a different way forward. The GLC was the most radical of these. And it was defeated by a similar combination of conservative (little c) forces: not only Thatcher and her troops, but also the bulk of the Parliamentary Labour Party, and those elements across the political spectrum that took it upon themselves to sneer at the attempts to develop a politics that was feminist, anti-racist and anti-homophobic, as well as challenging to capital. This GLC was more radical than the national Alternative Economic Strategy in particular along precisely those dimensions of openness and democracy the lack of which had vitiated the AES as a viable left alternative to neoliberalism. The viciousness of the attacks on that GLC, and the fact that these attacks continued long after its abolition, with the clear intent of destroying it even as an imaginative resource for the future, are themselves a hint of the potential it offered.[5] Certainly that history evokes the possibility that cities (or, even, more generally a certain kind of radicalised politics of place) could be crucial not only to neoliberalism (as is now so frequently averred) but also to the potential imagination of and experiment with an alternative. Maybe they are indeed what Sassen has evoked as 'a strategic terrain': 'Major cities have emerged as a strategic site not only for global capital but also for the transnationalization of labor and the formation of transnational identities. In this regard they are a site for new types of political operations' (1999, p. 189). It was in London, too, that took place Rupert Murdoch's destruction of the print unions: 'one of the most symbolic events of the Thatcher decade' (Lansley, 2006, p. 157). And out of the rubble of the battle over the future of London's docklands rose Canary Wharf: 'probably the best symbol of London's recent development as a financial centre' (Larsen, 2006).

*   *   *

When Ken Livingstone won the election to be mayor of London on that day in 2000, roundly and cheerfully defeating the New Labour power-brokers, he began his acceptance speech with words that brought immediate laughter: 'As I was saying when I was so rudely interrupted, fourteen years ago . . . '. The good times, the old creative insubordination, it seemed on that bright morning, were back.

But those fourteen years had vastly changed the political and economic context within which any London government could operate. And New Labour, both terrified at the potential for a reinvigorated democratic radicalism in the capital city and also committed to the prosecution of a basically neoliberal agenda, had consented only to a new Authority (the GLA) with strictly curtailed powers and a very different financial basis. The 1980s GLC had been located at a decisive moment, Tory government still in power and neoliberalism not yet consolidated. The current GLA finds itself facing New Labour in an era when neoliberalism is hegemonic.

Moreover, over this same period the left more widely, and partly in response to these changes, had itself been evolving its analysis of the possibilities for intervention at the level of the local state. When inner-city problems first came on to the political agenda in the 1970s the characteristic analysis promulgated in government circles was that there must be something wrong with the cities. They were characterised (caricatured) as full of the unemployable, the unskilled, single parents and generally those 'left behind' when all sensible normal people had fled to the suburbs. Quite apart from the empirical inaccuracies of this, it was – we argued – a geographical version of the more general political tendency to 'blame the victim' for the problems they were facing: cities had failed in the competition for jobs and thus what was needed was area-based policies to improve their potential. The answer to this by the left and by progressive intellectuals was to reverse the terms of the argument: it was not the inner cities that had failed capitalism but capitalism that had failed the inner cities (see the work of the Community Development Projects; and Massey and Meegan, 1978); the cities were at the sharp end of a more general process of deindustrialisation; in consequence it was no good having policies based only at the urban level

– wider and more systemic changes were needed at national level too.

So when the new urban left gained control of municipal councils in the early 1980s and set out to challenge Thatcher there was, therefore, a puzzle. Was there now more possibility of local intervention? There were a number of elements in the response to this. First, much of the politics adopted in the cities, and especially in the GLC, was exemplary and rhetorical. The aim was to argue for alternatives and to establish through small and symbolic interventions the fact that an alternative politics was imaginable. In other words, if it was not possible with the powers and resources at hand fully to address the problems of the cities, nonetheless the possibility in principle of doing so could be established. This, then, was a politics which was also addressed to the world beyond the cities themselves. Second, of course, such a strategy was particularly important and effective because it was directed against a national government which was Conservative. London and the other cities stood, effectively, as a voice against the dominant national politics. Third, nonetheless, there was also an analysis, particularly in the GLC, which attempted to establish the possibility of effective intervention at local level. Here the argument was that capitalism itself was changing, away from the cost-sensitive mass production of Fordism towards smaller-batch and higher quality production. This was argued to be true particularly in 'First World' countries and in their cities. Moreover such production focused on quality and skill rather than only on price. Maybe, then, there was room for manoeuvre for improving the conditions of inner-city labour while still remaining competitive – in other words a less degrading strategy for working-class jobs than that on offer from the Thatcher government. An enormous programme was set in train to explore these possibilities and to work out a strategy of 'restructuring for labour' (as opposed to restructuring for capital). The published documents, *The London industrial strategy*, *The London financial strategy*, and *The London labour strategy*, stand as a monument to this inventive period.

Since then, the possibilities for local intervention have shifted once again. On the one hand, they seem even more restrained. 'Globalisation', as rhetoric and as form of economic organisation, has reinforced the pressure on each local place to compete with all others (see, for instance, Harvey, 1989). In that sense, the room for local manoeuvre seems only further reduced, the pressure to engineer conditions attractive to mobile capital even more intense. As Peck and Tickell put it, there has been a neoliberalisation of inter-local relations – 'a more nebulous and a more daunting adversary' (2002, p. 387). On the other hand, with the movement from roll-back to roll-out, London is less dominantly shattered by decline; it feels more obviously a centre of the new economic age. 'Globalisation' comes to be experienced as less of a negative force. Indeed it raises the question – or should raise the question – of the geographical imaginary of globalisation itself. Those who would have us bend the knee before its inevitability evoke it as a place-less force; yet it only exists and is reproduced, and requires constant maintenance and modification, through locally situated processes. In that sense, the local and the global are, as the mantra has it, mutually constitutive. There is thus some purchase, at the local level, on so-called wider mechanisms; some possibility – in principle – for active local intervention. The left analysis in the 1970s assumed that the local was a product, indeed a victim, of the global. What is clearer now is that the local is, rather, also one of the moments through which the global is constituted. However, in the context of sharply uneven development, the degree and nature of purchase on the constitutive social relations, and thus the possibilities for local intervention, will vary dramatically between places. In the case of London, claiming simple victimhood should be out of the question; London is, rather, a site of the generation of some of the processes, relations and conventions to which it is now itself also subject. The financial City is itself one of the sources of that common sense that insists that places must compete with each other (those extra-local spatialities referred to above). In London, the question of resisting neoliberal globalisation returns as a local question too.

Such a reimagination poses to 'London', as a site of power within the networked relations of globalisation, questions of its own responsibility. It is one of the ways in which is raised that question: what does this place stand for? And that question of responsibility is posed along three dimensions. There is the responsibility that inheres in its role as international world city and that raises questions about its relations with the world beyond the United Kingdom. This is addressed in Part III. There is the responsibility that pertains to its role as national capital, and that must address the increasing dominance of this city-region within the country as a whole. This is the focus of Part II. And there is the question of the sources and implications of the increasing inequality that resides within the city itself. This is the arena on which most analyses concentrate (whereas my aim is to begin to imagine a politics of place *beyond* place), and it will therefore be treated more briefly here.

*   *   *

The rest of this chapter looks briefly at the last of these. Livingstone's stated aim is to work towards making London into an exemplary, sustainable world city. And the range and reach of the ideas, plans and studies that have been produced under his tenure is impressive indeed. In a host of ways this is an administration that is different, making an attempt to work out an imaginative, and progressive, politics of place. There is a large literature on the left which imagines neoliberalism as a kind of irresistible force, and understands all national and local politicians as virtually automatic transmission belts for its implementation. This is not so in London. Indeed, at a more parochial level, Tony Blair's scheme for city mayors was precisely envisaged as a mechanism whereby the national politics of New Labour could be directly piped into local areas. It did not work – especially not in London.

The wide range of studies and plans produced by the GLA gives witness to a strong commitment to combatting inequality along numerous cross-cutting dimensions, including those less often attended to, such as age, disability, the independent mobility of children. There are progressive policies on wages paid to the authority's own workers. There is an energy strategy, a food strategy, a walking plan for the city, a set of proposals for the rivers and lakes – with an emphasis on democratic access. This is a set of documents that deserves a wider readership than it will probably ever get.[6] However, the main economic aim that underlies it all is to support the continuation of the existing growth of financial and business services. It is these that are seen as the 'chief engine of economic growth and job-creation' (GLA, 2004b, p. 8). It is here that is most clearly demonstrated the difference from the 1980s, when the future was more readily understood to be open. Now, in the twenty-first century, 'The London Plan cannot realistically reverse these strong, deep-rooted factors driving change, nor does the Mayor wish it to do so' (ibid., p. 3). This, then, accepts and works with the now established neoliberal hegemony. So too does the emphasis in the economic strategies on the need to attract inward investment, and the thread, which runs through, of the need to compete with other places. Indeed that other face of London as world city – its cultural mixity – is sometimes mobilised in this competition with other places. This happened in the competition for the Olympics, and cultural openness has been proffered as an attraction to potential Chinese inward investment (Livingstone, 2006).[7] In the first, draft, version of the plan, which was put out to wide consultation, these themes were dominant: the City and business services were the single central focus of growth. (This meant, too, that growth would be concentrated spatially.) It was this focus that proved to be most controversial. From across the political spectrum there was criticism of the emphasis on finance and business services alone. The resulting revised plan (GLA, 2004b) responded to this with a greater emphasis on promoting a diverse economy and on recognising the significance of London's town centres and suburbs that lie beyond the central core. There was no pulling back from support for finance and business services, but other things were added in – there was more recognition of 'the ordinary city'.

Once again, part of what is at issue here is a geographical imaginary: an implicit geography that organises our social understandings, that supports – though usually without being mentioned explicitly at all – the analyses in documents (as well as practices) of all kinds. At the extreme, an understanding of the London economy as overwhelmingly dependent upon finance and business services holds out a vision of the city as a kind of pyramid with those sectors on its heights: a shining centrepiece. Benefits from these central sectors are then imagined to flow down and outwards to the rest of us, through direct engagement or employment within those sectors, or through a range of multiplier effects. Here space is a smooth surface across which benefits flow, to passive recipients. It is the perfect geographical underpinning for trickle-down economics. The only autonomous energy is that at the centre. This is the geographical imagination that accompanies the synecdochal characterisation of cities examined in chapter 1. The part (here finance and business services) is made to stand in for the whole: there is recognition of only one active agency. It is an imagination of the city that can be deeply influential on policy, focusing it too on only this one element, obscuring the needs and potential of the place's wider character (Amin, Massey and Thrift, 2000; Robinson, 2002). The imaginative geography of the draft *London plan* was never quite as stark as this (though we shall see in Part II that some of the geographical imaginations held in the City and associated sectors are indeed precisely of this form). But it was nonetheless to this broad proposition that those participating in the consultation process reacted so strongly. This was a refusal that all should be subordinated to relations focused on the financial core; it was an assertion of other autonomous lives, sectors, voices. It was, implicitly, an assertion of the multiplicities of space (Massey, 2005). The response, as we have seen, was a genuine attempt to acknowledge this greater diversity. It did not perhaps go as far as Buck et al. in recognising diversity as the crowning characteristic of the capital, but it attempted to go some way in that direction.

'Diversity' itself, however, implies a geographical imaginary of a simple plurality, a harmonious juxtapositioning. It too can be inadequate if it fails to recognise (just as in the case of some conceptualisations of multiculturalism) the relations that exist 'between' the elements of this diversity, relations which mutually construct the different elements, that perhaps set them in conflict. This is an aspect of the city, and of urban space, that is missed too by those simply celebratory depictions of city life. There is indeed much to 'delight' in, but the tensions within the diversity can all too easily be ignored. The victory of neoliberalism over any alternative more democratic, more egalitarian, future and the associated victory of banking, finance and related sectors and of a vision of London's status as this particular kind of world city has changed the conditions of existence of all else. As was argued in the last chapter, other parts of the economy find it more difficult to prosper, even survive, under this newly established hegemony. Poor people are pushed further into poverty. This is not a simple diversity, but a multiplicity riven with tensions. Urban space is *relational*, not a mosaic of simply juxtaposed differences. This place, as many places, has to be conceptualised, not as a simple diversity, but as a meeting-place, of jostling, potentially conflicting, trajectories. It is set within, and internally constituted through, complex geometries of differential power. This implies an identity that is, internally, fractured and multiple. Such an understanding of place requires that conflicts are recognised, that positions are taken and that (political) choices are made. Not all sectors of the economy can be equally encouraged; their interests cut across each other. It is an understanding that requires too a more rigorous approach to redistribution. Real redistribution cannot take place without addressing the relations within the multiplicities of the city. For it is through those relations that are in part produced the very inequalities that redistribution is designed to address. The *London plan* charts in many ways the reverse 'redistribution' that has taken place over recent decades. Perhaps most indicatively: 'In 1980, the top 10 per cent of full-time male earners in London had weekly earnings just over twice as high as those in the bottom 10 per cent. In 2000, the ratio had grown to nearly four times' (GLA, 2004b, p. 32). In less than twenty-five years this inequality had doubled. It cannot be tackled by addressing only the lowest earners. For what happens to them is inextricably related to that 'top 10 per cent of full-time male earners' that lie at the heart of the problem.

The difficulties that are thrown up by not addressing this issue directly are evident in some of the policies already devised (by national government, local government, trades unions, and others) to respond to the problems London and some Londoners are facing. One set of responses resorts to what might be called 'special measures' to enable the city to continue functioning. Such measures include top-up wages for Londoners over and above wages in other regions ('London weighting'), a whole raft of policies to attract and retain 'key workers', and a major programme of 'affordable housing' by the GLA. This last includes social housing, intermediate housing and, in some cases, low-cost market housing (see GLA, 2004b, chapter 3A). The need for such a large and high-profile intervention is seen to result from the absolute growth of the city, from the difficulties of keeping the city working, and from the acknowledged two-way inter-action between the shortage and high price of housing on the one hand and the increasing inequality in the city on the other. In response to this, a strong set of policies on the pro-vision of such housing is set out. In its determination to address the question of housing poverty this is far ahead of anything proposed by national government.

There is one observation that should be made about this immediately. Smith among others has made a strong argu-ment that gentrification has become a significant neoliberal urban strategy. He writes of 'the generalization of gentrifica-tion as a global urban strategy' (2002, p. 437). It is certainly the case that the gentrification of central urban areas since the end of the 1980s is quite unlike anything that could have been foreseen in the dark days of inner-city decay. It is also the case that this is a class victory that has gone hand in hand with the establishment of neoliberal hegemony: 'the new phase of gentrification . . . dovetails with a larger class con-quest, not only of national power but of urban policy' (ibid., pp. 440–1). And it is mirrored in a host of urban strategies around the world. It is also a conquest that installs property capital as a central motor of urban 'regeneration', and which sees such regeneration/professionalisation as crucial to inter-urban competition, itself an aspect of the neoliberalisation of spatial relations. These are all important points. It is not clear, however, that in the case of London gentrification as such can be said, at the present moment, to be an explicit 'strategy'. Once again it is important not to overgeneralise, for this will not only fail to register the differences between places but also, and more importantly, erase the possibility for, and recognition of, political attempts to act otherwise (see Larner, 2003; Ward and Jonas, 2004). The genuine strength of the affordable-housing policies in London is one such attempt.

Nonetheless, it is undoubtedly the case that the unre-served support in London for finance and the surrounding constellation of sectors as the centrepiece of economic strat-egy does effectively buttress the further gentrified takeover of the city. Moreover, given this context, compensating strategies such as the provision of affordable housing are unlikely to work as well as intended. Given that they do not address the underlying tension, they become quickly over-taken. 'Affordable' housing gets caught up in the constant escalation of prices. Indeed, precisely because they respond only to immediate symptoms, soothing temporarily the problems of this kind of growth, they contribute to the per-petuation of the long-term dynamic that is at the root of it all. At worst they are a patch-and-mend approach; they do not address the production in the first place of the problem.

Again geographical imaginaries play a part. On the one hand, that view of a whole range of 'global' forces as some-how always originating outside of the place (city, nation) in question is not only analytically untenable (globalisation is

produced in places), it is also, though not always intention-
ally, a means of avoiding responsibility. When the *London
plan*, on its opening page, asserts that 'Over the past 20 years
London has changed dramatically. Some of these changes
are being driven by international forces, including: the globalisation of many economic sectors, and the dominance of
the finance and business sectors, frequently interlinked with
dramatic advances in technology . . . ' (GLA, 2004b, p. 1), it
is both correct (these forces are indeed internationalised) and
evasive ('London' had no small hand in these developments).
On the other hand, the establishment over the last three
decades of the hegemony of neoliberalism, and of the
London/South-East-based sectors and social forces in which
it is based, has also been a clear project of successive *national*
governments. The importation of the US model of astro-
nomical remuneration packages is at the heart of London's
problems of poverty and inequality, yet it is at the level of
national government that the explicit condoning of fat-cattery has taken place. 'London', then, is hemmed in, in a kind
of spatial trap, as both generator and beneficiary of, and suf-
fering from, the sharp exacerbation of national (and international) levels of inequality.

The London government's ability to address this inequality is also constrained by a lack of powers and a lack of
resources. Indeed, Gordon (2004) suggests that these constraints are the impetus behind the presentation of London
as 'a global city'. On this argument, in spite of all the caveats
that must be registered about the reality of global functions
being either the major element in London's economic base
or even the main 'driver' underlying the city's growth since
the 1980s, it has been seen as necessary when bargaining for
more resources for what is in fact the richest city/region in
the nation for its representatives to insist that 'growth in
London is different in kind from that in other UK cities, and

hence not competing with them, only with foreign cities; and that additional infrastructure investment is required in London in order to sustain growth in the goose which lays the national golden eggs' (Gordon, 2004, p. 12).

The geographical imaginary of the goose that lays the golden eggs will be explored in Part II. What this argument does raise, however, is the question of why, instead of – or even maybe as well as – clamouring for more resources from the national pot, London's potentially powerful voice is not raised against the nationally deepening levels of inequality, and in particular the shameless greed of 'the 10 per cent of top male earners', which cause such problems for it.

London is entangled in a web of spatialities that can be addressed politically in different ways. There is some choice here. Moreover, addressing issues of inequality does not have to take the form of post-hoc redistribution. The greatest degree of maldistribution, of the production of inequality in the first place, is effected by neoliberal economic strategies. Here, both national government and London government play a role. 'London' is not helping itself in continuing to prioritise its neoliberal constellation. It is, rather, participating in the persistence of its own inequalities and the further securing of the national neoliberal hegemony and the lives of the super-rich. In this, New Labour at national level has taken the lead. As Labour MP John Denham (writing of national policy) has written: 'Rather than attempt to extend elements of social justice into the private sector economy we have preferred to run social justice and the market as parallel activities, relying on the free-market economy to provide the wealth from which the social justice can be created' (Denham, 2004, p. 73, cited in McIvor, 2005, p. 85).[8]

One thing that this means is that the contest in and over London (at present subdued) really matters. And that contest involves a politics actively engaged in by far more than the local state. The national government, spokespeople for the North, trades unions, NGOs, grass-roots campaigns, the City and the Corporation of London, a plethora of consultants, and think-tanks and academics . . . all these and more are engaged in this contest. And it returns us again to the question of the potential for a radicalised politics of place – can such a politics reassert, indeed reinvent, itself in such different and apparently unpropitious times?

Part II

THE WORLD CITY
IN THE COUNTRY

4

THE GOLDEN GOOSE?

The growth and self-assertion of global cities has taken place in many countries in a national context that has become geographically more unequal. This may be because the urban in general has been prioritised over the rural; it may be that the need to compete globally on the world stage forces countries to put in the shop window, as it were, their most internationally appealing (which is to say Western) and already successful places – and thus success on the international stage can entail increasing inequality *intra*-nationally; it may more simply reflect the clustering tendencies of the new global elite of corporate bosses, high-tech workers, and financial operators. Each case will be different. But there is no doubt that intra-national inequality has been increasing around the world. These are often violent geographies, yet we rarely get to vote about them directly. They are, nonetheless, present on the political agenda. Resentment in excluded rural areas, antipathies between rich regions and poor, are frequent. In one way or another these geographical inequalities have to be addressed, or legitimised. One question is how the booming regions get away with it.

This is the question the present chapter begins to explore. The argument turns on a number of interwoven issues. There is the discursive dominance of these global cities (their voices, precisely because of who is there, are often louder). There is the extension of neoliberal economic thinking to inter-regional relations, which dragoons non-dominant regions into line. And there are the powerful geographical imaginaries embedded within even the most apparently mundane of documents.

There is no question but that the reinvention of London as a global city has been set within a national geography that is markedly, and increasingly, unequal. The Institute for Public Policy Research (IPPR) is adamant:

> The 'North' is poorer than the 'South'. Although some dispute the existence of a 'north–south gap' in prosperity, it is clear that within the UK there is a 'winner's circle' in the Greater South East of the UK (consisting of London, the East of England and South East regions and parts of the South West). The rest of the country has lower levels of prosperity, and three regions lag significantly: Northern Ireland, Wales and the North East of England. (Adams, Robinson and Vigor, 2003, p. i)[1]

The IPPR study also points out that this inequality has been growing since the early 1990s. The European Commission (1999), in its *Spatial development perspective*, demonstrated that, in terms of GDP per capita, the UK has the greatest degree of regional inequality in Europe and that the main factor in this inequality is the dominance of London. More recently, Dorling and Thomas, after their exhaustive analysis of the 2001 Census in the UK, come to a single overall conclusion:

> Our conclusion is that the country is being split in half. To the South is the *metropolis of Greater London*, which now extends across all of southern England in its immediate spatial impact. To the North and West is the *archipelago of the provinces*, a series of poorly connected city cluster islands that appear to be slowly sinking demographically, socially and economically. (2004, p. 7, emphases in original)

And most recently, a research report from the think-tank Reform began with the words:

> Regional economic performance has become increasingly imbalanced over the last ten years. A series of key indicators shows a clear and growing difference between dynamic regions, in particular London, the East, and the South East, and challenged regions such as the North East, North West, Scotland and Northern Ireland. (Bosanquet, Cumming and Haldenby, 2006, p. 4)

It is a situation of which everyone is aware. This includes a Labour government that, although in its New-ness it might find the circles of London's elite more socially compatible, nonetheless still has its electoral base among those who have suffered most from the collapse of the old class settlement. The awareness extends also to those who are benefiting most from the new London-centred inequality – and who frequently find it necessary to explain to the rest of the country why and how the widening inequality between regions can be beneficial to all. In keeping with their different structural positions, the arguments put forward by these two groups (New Labour national government and London-centred private capital) do differ somewhat, as will be seen. But what they share is an understanding of London/the South-East as the golden goose of the national economy. For such a position to be tenable it requires a very particular background geographical imagination. Most generally it requires an assumption that London's growth and relative wealth is at least benign, possibly beneficial, or in the most extreme formulation essential, to the economic health of the rest of the country. And the key way in which that has been figured is to argue that London, in the unappealing terminology of the new managerialism, is (or is the location of) the 'single driver' of the national economy. As the golden goose it lays eggs for everyone.

New Labour, coming to power after Thatcher had 'destroyed the old historic bloc and created the basis for a new neoliberal era', accepted the mantle and set about its mission of consolidating the new settlement. Crucially, it needed to do that in 'the regions' (and in the cities); the task was a clearly spatialised one. And Labour, given its own geography, was clearly the party to do it. While on the one hand its mission was to run with the victorious forces based in London and the South-East, on the other hand it did need to respond to the evident geographical inequality affecting many of the very people who had voted it into power.

The strategy that emerged *combined* the recognition of regional inequality and the need 'to do something about it' with the economics and cultural norms of the neoliberalism emanating from London/the South-East. It interpreted the former in terms of the latter. The demands from the regions were taken up and transformed through the dominant logic. In this, the approach to regional inequality was entirely consistent with the mode of 'double-shuffle' that has characterised the whole New Labour project – 'the social-democratic route to neoliberalism' (Hall, 2003, p. 20).

Spreading the culture of neoliberalism to 'the regions' and into the specific policy area of regional inequality did not proceed so much by overt ideological propaganda aimed directly at conversion to entrepreneurial values. Rather what was altered was the environment, the rules of the game, through which people were constrained to operate. 'You change what individuals do not by changing their minds but by changing their practices, and thus the "culture"' (ibid., p. 18; see also Peck, 2003; Hudson, 2006b). That indeed was part of the geographical extension of the new hegemony.[2] But, if this was the strategy of New Labour, it was also supported by the continuous production of documents, statements, arguments and pronouncements, by the resurgent elite itself. By this means, it was sought to establish a new 'common sense' in relation to a regional problem that was now more than anything the problem of the dominance of London.

A basic position was crafted thus: there is indeed regional inequality, and it must be addressed; however, this must be done by analysing and correcting the weaknesses of the 'regional' economies and/or by helping them find the means to improve their competitive position. What is certain – it is insisted upon again and again – is that encouragement to 'the regions' must in no way be allowed to challenge, question, or in any way restrain the growth in London and the South-East of England. Thus, in perhaps the most measured formulation of this position, Her Majesty's Treasury, in a joint document with the Department of Trade and Industry, argued that 'attempts to address regional differentials must be done by a process of levelling-up, not levelling down. . . . whilst regional economic policy must aim to strengthen the indigenous growth potential of all regions, the focus should be on the weakest regions, without constraining growth in the strongest' (HM Treasury/DTI, 2001, p. 43). The hegemonising effect of this discourse can be detected in the fact that even those arguing the case for *northern* cities seem on occasions constrained to follow the line. Thus, at a Department of Trade and Industry conference on 'London and the Rest of the UK – the economic relationship', convened with the Regional Studies Association, a spokesperson from a northern council, who presented an excellent case for a more regionally egalitarian country, nonetheless performed the ritual genuflection to the untouchability of London: 'The London economy is a huge asset to the UK', before adding, cannily, 'But we worry that it will overheat and lose out to other global cities' (www.regional-studies-assoc.ac.uk). The discourses surrounding 'The Northern Way' and 'Core Cities' initiatives, both of them attempts to generate growth beyond the South-East, also often acquiesce in this imagination, explicitly positioning these policies as supportive of London in its expansion, for instance by acting as a safety-valve to avoid overheating. In his explanation of 'the Government's plans to close the productivity gap between the English regions', Alan Johnson, as government minister for manufacturing, wrote:

> How are we going to do it? There can be no question of holding back London and the South East. They must grow in order to tackle poverty in parts of London, as well as in Hastings, the Isle of Wight, Brighton and Norwich.
>
> Thankfully we've moved on from the one-dimensional debate about the North–South divide. We must aim to strengthen the indigenous growth potential of all regions, the focus must be on the weakest regions, without constraining growth in the strongest. (Johnson, 2002)

This pairing of a casual and unargued disparagement of a regional policy of previous decades with an absolute insistence on the importance of in no way challenging London and the South-East has now become part of an established mantra.[3] It is clearly important in serving the interests of the neoliberal elite based in that corner of the country.

It is also a view that has a direct parallel in the debate about national levels of poverty and inequality. New Labour have been quite forthright in both rhetoric and actions around the issue of poverty, particularly in its most severe forms, but they have been extremely reluctant to address inequality (see the discussion in Jackson and Segal, 2004, for instance). What has lain behind this is a refusal to address the question of wealth – the enormous growth in riches of those at the highest levels of income. The issue of 'fat cats' is not one that has disturbed them. As Peter Mandelson, right-hand man to Tony Blair, is reported as having boasted in a speech to Silicon Valley executives, the Labour Party is now 'intensely relaxed about people getting filthy rich' (cited ibid., p. 7, quoting Rawnsley, 2001, p. 213). We have seen in chapter 2 some of the effects of this within London – in other words the working out of the dynamics of inequality *within a particular place*. The unwillingness to address the wealth of London/the South-East in the context of the inequality between regions is the *inter-regional* equivalent of this more general stance. Just as 'the demands of a global elite' render them 'untouchable' (Lawson, 2006), so too in the geographical arena London and the South-East – the home-base of this elite – can in no way be challenged or restrained. As we shall see, however, the dynamics of inequality between regions, although different from those operating within a particular place, are equally harmful.

The powerful geographical imaginaries, the implicit con-ceptualisations of space, that enable such positions to be maintained can be unearthed by examining the government's formulation of the issue. Government documents are not, perhaps, the most exciting foci of analysis. Yet they are vital.

It is, in part, here that takes place the working-out of the new rules and logics that will be rolled out to mould behaviour and, eventually perhaps, thinking. Laying bare what these documents are doing is necessary for combatting the new 'common sense', for pointing to the fact that it is not common sense at all.

In November 2001, the Treasury and the DTI published an enquiry into 'the regional dimension' of 'productivity in the UK'. In itself this was a very positive move – an explicit recognition that the national economy is not a spatially undifferentiated entity and that regional inequality should be addressed. However, the resource the Treasury and the DTI drew upon in pursuit of this enquiry was economics, and economics of a particular hue. This was not unreasonable in that the aim was to set out 'the economic analysis that underlies the Government's approach to regional economic policy' (HM Treasury/DTI, 2001, p. v). But it immediately brought in its train a very particular way, also, of thinking about space. Indeed, as Martin has so ably argued (Martin, 1999), it might rather be called a way of *not* thinking about space. The particular form of economic theory adopted, and the (implicit) conceptualisation of space, were integrally related.

An insight into this mind-set can be derived from an opening statement of position:

> The high degree of persistence of regional differentials points to significant problems in under-performing regions. Growth theory would suggest that market forces should result in convergence of regional GDP per capita over time. A lack of convergence in the medium to long term is therefore a likely indicator of serious market failures in a number of UK regions or countries. (HM Treasury/DTI, p. 3)

A number of interlinked aspects of the overall framing imagination can be detected at work here. All of them reinforce a
neoliberalisation of the regional question.

First, there is a foundational belief in equilibrating market
mechanisms (the 'lack of convergence' *could* have indicated,
rather, a serious *theoretical* failure, but this is not considered).
The assumption of the benefits of market mechanisms and
the need to facilitate them runs through the document. In
each analysis, any failure of convergence is traced to market
failure or else to 'coordination failure' preventing agglomer-
ation in poorer regions. The document's basic assumption is
that markets, if not interfered with, will lead to greater equality. As Dunford crisply puts it: 'According to market funda-
mentalists, market systems give rise to processes of catch-up
as less developed areas grow more quickly than more developed areas. Empirical evidence does not lend a great deal of
support to this thesis, nor indeed do recent developments in
convergence theory' (2005, p. 163; see also Dunford, 2003).
Other arguments that point to the existence of other forces
are not entertained by this theoretical position. More
unnerving yet is that at the end of the Treasury/DTI document these assumptions are re-rehearsed as a *finding*: 'The
analysis . . . identified market and coordination failures as the
likely cause of the shortfall in the productive potential of
localities, countries and regions in the UK' (HM Treas-
ury/DTI, 2001, p. 44).

Second, to make this model work there must be identifiable
regional economies. The most baffling assumption that
underlies the document is that regional economies are separate from each other. They perform, they 'under-perform',
they succeed. This is crucial. No serious account is taken of
relations between regions; space is not in any way conceptualised relationally. Regions are always already given to the
analysis. They exist and then they compete. It is the

Newtonian, billiard-ball world – here of isolated regions instead of isolated individuals; but it does not reflect the real economy. And it makes it very difficult to entertain the possibility that the growth in one region might have deleterious effects on the prospects of others. This symptomatic inattention shows up clearly, in a rather different way, in the document's outline of 'the Government's approach' (HM Treasury/DTI, 2001, p. vi). There are two key elements in this approach. First the national government is to aim to create a stable macro-economy; and second it will build the capacity of regional and local institutions. But this is to ignore the possibility both that arriving at a stable macro-economy might mean responding differentially to signals from different regions and that the measures put in place to build a stable macro-economy might have differential effects between regions. (This, indeed, has been a point frequently made about the effects of regional inequality – see chapter 5.) In other words it ignores the implications at national level of existing regional inequality.

Third, the economy is considered as a world unto itself. And even within the economic, no consideration is given to power in any of its forms. More broadly, nowhere in the document does the fact, for instance, that London is the capital city and that this might have effects receive any serious attention. The authors confront in total bafflement the fact that the USA and Germany have a greater number of thriving centres within their national boundaries. Although it is clearly the case that the economic mechanisms that they do consider must have a significant role to play, the failure to see beyond that abstract economic landscape leaves them with an impoverished geography in which a whole range of other mechanisms and relations – and thus a whole range of potential policy instruments – is unavailable. In Amin, Massey and Thrift (2003), we argue not only that it is necessary to conceptualise the national geography in terms of relational space but also that this geography of relations is importantly constructed through relations of power, including political and cultural, as well as economic, power. The highly centralised geometry of power in the UK has effects on the potential of all constituent regions: it is a crucial force in the production of regional economic inequality (see chapter 5). It engenders a colonial relationship of many dimensions between London and the wider South-East on the one hand and 'the Rest of the Country' on the other, it induces a persistent bias towards the South-East in a range of purportedly national policies, it entails an assumption that all important national icons must be located in the capital (an equation of 'nation' with London), and it reinforces, through spatial concentration, a tight-knit, elite, class structure that revolves endlessly around an area of a few square miles. New Labour's endorsement of that class has been apparent. Such arguments are contestable – and have been contested – but the Treasury/DTI document, in its own particular imaginative geography, does not even allow them to be opened up to consideration.

Here, then, regions essentially do it on their own. It is a moral geography that mirrors the government's 'social exclusion' agenda, with 'the regions' being defined in terms similar to the socially excluded: what is needed, the argument runs, is remoralisation and, if only the 'socially excluded' regions would sort themselves out (become more entrepreneurial or competitive), they could join the mainstream. The criteria by which they are judged constitute London and the South-East as the norm, or model. This is the neoliberalisation of inter-regional relations. It is one of the means by which the regions beyond London/the South-East are bound into the terms of the new hegemony. A language of individualised 'performance' permeates the document: 'the weakest regions' (HM Treasury/DTI, 2001, p. 1), 'London has been able to achieve' (p. 9), 'poorly performing regions' (p. 44) and so on. The rhetoric of London succeeding by its own efforts is relevant here too, for it is a moral rhetoric like the rhetoric of social exclusion. This is not to argue that such characterisations are without any relevance at all. The 'performance' of regional economies indubitably depends on internal resources and how they are mobilised, as well as on the relational tissue, the power-geometries, within which the region is set. Both aspects need to be brought into account. But it is also the case that characteristics such as 'weakness' and 'strength' are specific to particular space-times; at different historical moments and under different politico-economic ideologies their meaning and relevance will vary. One is inclined to ask in what sense this is a *regional* problem, or a *regional* success, but this teasing out of the complexity of forces is not an issue that can be addressed in the framework adopted in the document. The spectacularly renewed growth of London over recent decades has been precisely about setting new terms which reflect the interests of the elite and of their base in London/the South-East.

An even more startling analysis was produced in 2004 by Oxford Economic Forecasting for the Corporation of London (local authority to the financial City). Here, in contrast to the analysis by the Treasury and the DTI, there is no political obligation to the regions, though there is a degree of embarrassment and an evidently felt need to legitimise. There is also a much stronger mobilisation of the theme of London as world city. A further, and fascinating, difference from the Treasury/DTI document is that there is here a serious recognition of the relationality of space, of the links between regions – indeed the study is entitled *London's linkages with the rest of the UK*. It is precisely the nature of that linkage that has generated the need for this defensive response. In his foreword, Michael Snyder, chairman of the Policy and Resources Committee of the Corporation of London, is clear about what is at stake:

> To many of us who live and work in London, the nation's capital is very clearly a valuable asset for the United Kingdom as a whole. A World City that serves as an engine of growth for Britain and a magnet for talent and enterprise on a global stage, London is a success story that we believe to be worth further investment. From other perspectives, however, the picture is often less clear. It is sometimes argued that London has grown at the expense of other parts of the country, and even that Britain would be better off if funds were diverted away from the capital to other regions.
>
> It is against this background that the Corporation decided to commission Oxford Economic Forecasting to undertake an analysis of the economic linkages between London and the rest of the UK. The mandate was left deliberately broad and open-ended. We already know, from work carried out by a number of consultants, that London makes a significant net financial contribution to the UK Economy, so the object here has been to explore the wider links and relationships that tie London into the country as a whole. (OEF, 2004, p. 4)

Their conclusions, summarised in their next paragraph, are as follows:

> Perhaps the most important conclusion is that London's growth has not been, and is not likely to be, at the expense of the rest of the UK. Rather than hindering the development of other regions, London's success has contributed to and stimulated growth elsewhere. The linkages between the capital and the country are such that economic specialisation

works to the benefit of all, while London's dual roles as a
World City and the premier international financial centre
add an extra impetus that few other capitals elsewhere in the
world can match. (Ibid.)

In other words, yes there are relations but they are benign, if
not positively beneficial. The way in which this conclusion is
arrived at merits some attention.

Most strikingly, there are some startling reversals of what
might be called the 'normal' imagination. Thus, for instance,
the huge disparity in house prices between London and the
South-East on the one hand and the rest of the country on
the other is accepted. This is a fact that is usually interpreted
as set within a dynamic that exacerbates the inequality
between the two parts of the country, and that exemplifies –
and is felt as exemplifying – the colonial nature of the relationship. Not so here. In this study, 'Since London house
prices are significantly above the national average, people
moving out of London to other parts of the UK are often
able to release significant amounts of housing equity, which
can then support spending in the regions into which they
move' (OEF, 2004, p. 36). This counter-intuitive account
depends on a version of the escalator-region argument:
young people move into London, acquire property assets,
those assets increase in value, the people later in life sell these
assets, leave London, and spend in 'the regions'. There are
two things going on here that it is interesting to note. First,
rather than examining the dynamics of the production of
inequality (in this case in house prices), the document takes
this inequality as its starting point, and looks only at the
inter-regional relations that derive from it. But second, and as
a consequence of this, this is effectively an argument that
inequality is good. It is not just London's high house prices
that are beneficial, but also the differential between them and

prices in other regions. 'Indeed, in 2002 the typical (median) amount of housing equity owned by mortgage-holders in London was £133,500, twice the UK average of £52,700. . . . This suggests that people moving out of London to other parts of the UK will often be in a position to realise significant amounts of housing equity' (ibid.). In other words, high house prices in London *benefit* other regions. The chutzpah of this is breathtaking. The fact, for instance, that this invasion of moneyed Londoners might intersect with processes under way in those other regions (in particular making it difficult for the new generation of existing residents to buy locally) is ignored – yet the consequent rise in house prices there has become a significant political issue.

A similar feeling of being in a hall of mirrors is produced by the interpretation of various ways in which the national economy is concentrated into London and the South-East. Instead of the concentration (for instance of headquarters, or of management, or of media) being an element in the production and reproduction of inter-regional inequality, here it is taken as given and then interpreted as a source of London's largesse to the nation as a whole. Thus 'London's financial markets provide access to capital for firms based throughout the UK' (OEF, 2004, p. 8) – when there is considerable evidence that such spatial concentration can result in a lack of finance for firms far away from (and out of sight of) that centre (Mason and Harrison, 1999). Or again, the concentration of head-office, managerial and strategic jobs and functions in London is interpreted as beneficial to other regions through the dispersal to them of back-office jobs (OEF, 2004, pp. 6, 17 and 23 et seq.) – call centres are mentioned as a case in point (whereas what this concentration – considered as a *process* – has in fact meant has been the draining from 'the regions' of particular social strata and of higher-level strategic functions).

There is a serious argument behind these interpretations. At the broadest level it is that London's position as a world city (that is, as more than a standard capital city) gives it a range of functions and assets which it would not otherwise have, and through which in consequence it can benefit the rest of the country.[4] A first reservation, then, is that, as we have seen in chapter 1, the claim that London is a global city in itself needs finer specification. The recent growth of London and its region is not primarily a result of these global functions (even though it undoubtedly does perform them). If what the document is addressing is, as it claims, the effects of the *general growth* of London, then this equation with world citydom is quite simply unwarranted. London's *growth* is just as much a product of financialisation, deregulation and marketisation. It is these interests that are being defended behind the banner of London-global-city. Moreover, even given this, what is further disarming is the lack of attention to the possibility that the inter-regional dynamics into which these phenomena are inserted might be part of a wider reproduction and/or exacerbation of regional inequality. In other words, the structure of the argument is, first to equate regional inequality with the fact of London's being a world city (rather than a run-of-the-mill capital city) and second to take this as given and ask what benefits or disadvantages in consequence accrue to other regions.

This is certainly, and in contrast to the Treasury/DTI position, to recognise the fact of relations between regions, and to recognise that the fortunes of one might have impacts upon others. Yet again there are issues here about the manner in which regions, and indeed space itself, are conceptualised. The imagination here is of space as a simple surface across which the benefits of the golden goose of London spread out to the regions. It is, again, a geographical imaginary adequate to the trickle-down version of economics. It does not ask how these outflows from London/the South-East intersect with activities and dynamics already under way in other regions. In this understanding, these regions – as in the housing case – are in effect the passive recipients of largesse from London. Moreover, this view does not address the long-term dynamics of the *constitution* of regions in the production of uneven development. It does not ask what might have been the effects on other regions of London's *becoming* a world city: that that in itself might be implicated in the exacerbation of regional inequality. Indeed, to come full circle, as with New Labour, inequality is not in itself a cause for concern. And the existence of great wealth, in this case in one part of the country, is understood as being simply beneficial to those who do not have it. This was a study produced for and welcomed by, the Corporation of London, a body whose aims include protecting and promoting the interests of the City. It is a view shared by the London Chamber of Commerce and Industry: 'The London economy leads, supports and sustains the UK economy' (Hill, 2003, p. 16) and, in no uncertain terms, by the Centre for the Study of Financial Innovation, which, confirming as it does so the City's foundation in laissez-faire and deregulation, reports:

> The UK government has no idea how the City works. Its desire to legislate (via labour regulation, anti-discrimination laws, financial regulation, tax) has almost universally negative effects. It could kill the golden goose. (CSFI, 2003; cited ibid., p. 31)

5

AN ALTERNATIVE
REGIONAL GEOGRAPHY

The geographical imaginations outlined in the previous chapter are not prime movers. Neither, on the whole, are they adopted explicitly, as a means to bolster a case. They are *integral to* those cases, and to the political positioning and the lives of those who make them. They are quite viscerally experienced, and on a quotidian basis *practised*, imaginations. They also play different roles. In some cases, for instance for London business, the imaginations are clearly self-interested, or maybe self-justificatory (the imagination of the golden goose as a legitimation, the tale they tell themselves). In other cases, perhaps for the representatives of northern cities, there may be an element of bowing to the inevitable, of tactical obeisance to the untouchability of London, a way of establishing a bargaining position.

These geographical imaginations are also performative. They help to figure the terms of debate and the actions in which they are embedded. They are vital underpinnings of political positions. They mould the common sense. But they can also be challenged. Indeed mounting any kind of challenge to the political positions that they support should mean also taking on these implicit geographies and conceptualisations of space. In recent years, a combination of the stirrings of discontent and political unease, and a simple recognition through the pressure of the empirical world that the complacent story of 'London benefactor to the nation' cannot always be made to hold, has provoked the beginnings of a nervous enquiry. In spring 2003 an editorial in *The Guardian* wrestled with the contradictions of proposals to initiate huge housing programmes in the South-East:

> A cabinet stuffed with northern MPs signalled yesterday that it would not just protect the current serious north/south imbalance, but actually dramatically increase it. The one spectre threatening the great engine of Britain's economic growth – the shortage of affordable housing in the south-east – is to be robustly addressed. . . .
>
> Some of this makes sense . . .
>
> But . . . Does this really make sense when Britain is already one of the most economically unbalanced states in Europe. . . .
>
> Some will say you cannot buck the market. If people want to live in the south-east, they should be allowed to. But this new development is not market-driven, but subsidy-driven. (6 February 2003)

In early 2005 a gathering entitled 'Can Britain afford London's growth?' was convened by Queen Mary University of London Public Policy Seminars. Or again, Cohen (2004), at the end of a fierce critique of developments *within* London, sets his interpretation of the capital city into a dismal regional context: 'Readers outside London may be wondering why they should care. The reason for national alarm lies in the prospective tenants of Kengrad's towers. Government and mayor agree London's boom in financial services must be allowed to continue sucking in people from the regions.' And Elliott (2004), taking the Dorling and Thomas analysis into a wider frame, writes of 'The United Kingdom of London'. The issue is clearly coming on to the political agenda.

But these are small voices. It is 'London' that is discursively dominant, as indeed are most 'global' cities in their national arenas. It is from here that the dominant narratives are woven, narratives which, as was seen in the last chapter, confirm in their very structure the geography of the single dominant voice. London is the most studied city (including the wider region) in the country. In part this is because it has an extremely active municipal government and mayor. In part it is because the national government is seemingly so afraid to challenge the dominant interests within the city. But mainly it is because of the presence of those interests themselves – London is the headquarters of so many peak institutions, especially of capital. And those institutions endlessly commission research designed to plead 'London's' case. No other region can *begin* to compete in terms of voice.

Indeed, the very naming of the geography frames the debate in a particular way. The DTI/RSA (Regional Studies Association) conference was called 'London and the Rest of the UK'. The Rest of the UK has even been condensed into further invisibility in an acronym: RUK. When you talk about 'the regions' you mean those parts of the country beyond London and the South-East (are the latter not regions too?). It is a classic binary of dominance and subordination. The rest of the country is what is left after London has been defined. The Rest of the Country is not-London to London's London.

For this understanding of regional inequality to be challenged it needs to be contested by another geography. Such a geography would bring those 'other regions' back into view by recognising them in their own right as locations of their own trajectories and – like London too – as the constantly shifting products of the relations within which they are set. This would be a geography of trajectories with interests variously concordant and contending; a field of multiple and potentially conflicting interests. Remember that the New Labour view makes the assumption that maximising the growth rate of London and the South-East and of each other region individually maximises the national rate of growth (let us disregard for the moment the caveats expressed in chapter 2 about growth as the singular goal to be aimed at). Yet this is not necessarily the case: the growth of one region, or some regions, may have negative effects on the growth of others.

There is in fact a long history of regional antagonism in the United Kingdom, not only between the constituent nations but also between the regions of England, about precisely such issues. There have, for instance, been antagonisms between contesting class elements within capital. The episode of the repeal of the Corn Laws is the early classic: when landowners (who wished to keep corn prices high to protect their own production) failed before the demands of those elements of capital that wanted (for a variety of reasons, but in the name, presciently, of 'free trade') to import cheap grain from the new world. Conflicts between finance and production have also been prominent since that period. Although London in fact was home to a huge manufacturing sector, the city's 'voice' was dominated by finance. The Midlands and the North stood for 'production' (including manufacturing and, for instance, mining). It is probably 'the return to the Gold Standard' that is the iconic moment here, when the financial interests of the City, in terms of a dominant and stable currency, won out to the detriment of production, and at the cost of unemployment in the North.

Disputes over the exchange rate have been endemic for decades. In the 1960s and early 1970s, when the emphasis of policy was on manufacturing rather than on services and specifically on manufacturing rather than finance and banking (Jessop, 1979), the conflicts between manufacturing and finance were muscular and continuous, finding expression in arguments over policy, tussles between government departments, and high-profile dissatisfactions in manufacturing over its investment relationship with the City. These antagonisms too had a geographical dimension. So also have the historic battle lines between these two groups over interest rates (the City favouring high interest rates and production preferring low). Indeed, the very *fact* of regional inequality exacerbates the potential conflict, for the pressure to put up interest rates is more likely to be felt first in London and the South-East. As Oxford Economic Forecasting puts it: 'There is, however, a view that London (and, by extension, the South East) creates problems for other parts of the UK, particularly its more peripheral nations and regions. One oft-expressed concern is that the Bank of England has to keep interest rates relatively high in order to prevent inflationary consumer housing and credit booms in London, at the cost of lower investment and growth in the rest of the country' (OEF, 2004, p. 14). It is a sensitive issue. Elliott's (2004) diagnosis is pessimistic:

> If the economy is increasingly dependent on the City, policy and politics will be driven by the needs of that sector. The migration of people to the south will drive up property prices, but governments will be reluctant to use fiscal measures – such as capital gains tax on first residences – in case there is a backlash in the key electoral battleground to the south of a line from the Humber to the Severn estuary.[1]

The result, argues Elliott, is that the Bank of England 'is left to deal with an overheated property market by increasing interest rates. The higher cost of borrowing and the hardening of the exchange rate is detrimental to manufacturing, further exacerbating the north–south divide.'

This is a complex mix. The straightforward economic interests of different fractions of capital are entangled with the shifting map of potential electoral allegiances, and regional inequality itself is part of the dynamic and not merely its result. It is an interplay of trajectories – of class, culture, economics, politics – within which both the existing fact of spatial differentiation and inequality, and the lineaments of their reproduction, are crucial elements. The base of Labour's historical electoral power lies in the North, yet:

> under Labour, 700,000 jobs in manufacturing have gone, the north-south divide has widened, and the City wields more power and influence in Westminster and Whitehall than ever . . .
>
>   Ministers cross these people at their peril, because the financial sector creates jobs, generates wealth to create the demand for a range of support industries and turns in a hefty trade surplus. Despite its genuine attempts to construct a meatier regional policy, the government knows its place. Its role is to make Britain safe for the financiers and the rentiers. (Elliott, 2004)

This is a geography of power, and what it produces is an (unequal) geography of democracy. Elliott's concern is with relations between the City and national government, both of which have their headquarters in central London. But the issue is also a wider one. In Amin, Massey and Thrift (2003) we lay particular stress on the tight relation within the UK between geography and power more generally. One aspect of this is that there is a 'distinctive spatial grammar which lies at the heart of the unequal distribution of power' (2003, p. 8). At its heart is a very small part of central London in which are concentrated the main executive instruments of national government, the flurry of ancillary institutions that surrounds them, the political news media, the Houses of Commons and Lords, and – what helps knit the whole thing together – the mesh of institutions of general sociability: 'restaurants, bars, clubs . . . These institutions are important in their own right as both lubricants and means of gaining influence' (ibid., p. 11). It is within this self-referential circle that is determined not only what political slants and lines and implicit understandings will structure 'public' political debate, but also what even *counts* as politics, and as news. Moreover the power, and the repercussions, of this political hothouse are further reinforced by the spatial concentration also into central London of the peak organisations of most elements of the national economy. Sampson reports that 'Today the elite looks much more unified . . . Visiting Americans are surprised that most people they want to see can be found at a few clubs, dinner-parties or gatherings, without ever leaving a handful of postal districts in central London' (2004, p. 355). In London, too, are collected the headquarters of a vast panoply of institutions of all sorts, from the arts, through sports, to religious, to non-governmental organisations. They 'have' to be here.[2] Finally, London is the presumptive location of 'the national', and this too, as we shall see, has effects. Moreover the fact of the spatial concentration of this mutually celebratory elite reinforces its dominance. This is a space of introspection and self-absorption that is profoundly exclusionary of those who lie beyond.

The implications of this tight hierarchical spatial grammar are far-reaching. For one thing, it is in itself undemocratic. It structures the national geography in a way that is exclusionary (within its own circle), colonial (in its production and reproduction of subaltern relationships) and in consequence demeaning (in that it produces a supplicant relationship with the metropolis) (Amin, Massey and Thrift, 2003, p. 9). In brief, it is a geography that both reflects and reinforces the thinness of democracy nationally. It also reinforces an elite, and thus class inequality, nationally. The A/not A structure of the terminology of 'the regions' (meaning beyond London) and 'Rest of the UK' is imbricated here right inside the (geography of the) practices and relations of economy and polity. The 'other' regions do not have their own, equal, voices or recognised trajectories.

Moreover, this concentration of power and leverage has effects on the economy. It exacerbates the regional inequality over which it presides. The imagination of 'the nation', in sphere after sphere, is constructed through London. Those charged with turning around the fortunes of 'underperforming' regions find it difficult even to put themselves on the map. An economic adviser to one northern city council talks of 'the subliminal role of reputation' and of the difficulties of changing reputation in the face of national invisibility or, perhaps worse, caricature. Said one southern entrepreneur: 'I have no business reason to visit Northern Cities and, like most people, I look towards the sun for my leisure destinations – so what I know about the North comes from media images' (cited in Sharp, 2004). At this particular gathering (one convened by central government) a representative from a northern city had brought some photographs. They were great; but the effect was demeaning. So too is the fact that delegations from the regions of the North, to negotiate with the national government, have to travel to the heart of the region which is their strongest competitor. The very geography of the movement reinforces the supplicant positioning.

Or again, while Oxford Economic Forecasting are not wrong to argue that London's position as a global city entails that there is a greater range of financial instruments and opportunities available within the boundaries of the country, their spatial concentration does raise issues of economic democracy. For instance, the spatial concentration of financial services into a small corner of the country can *deprive* other regions (or, small and medium-sized companies *in* other regions) of access to investment. Particularly in the case of smaller investments and venture capital, the investor requires knowledge, a 'feel' for the company into which money might be trusted. From the centre of London those more qualitative and knowledgeable, and small, engagements are difficult to establish in, for example, the North-East of England.

Or yet again, the construction of national government policy has for some decades been pursued through a lens in which London and the South-East are in sharp focus in comparison with vague notions of what lies beyond. Policies are shaped to respond to what is best known. This has been argued in education, for instance (Smithers, 2005). The highlighting of 'the knowledge economy' reflects a south-eastern perspective (Amin, Massey and Thrift, 2003; Local Futures Group, 2003). Policies not designed to be spatial turn out to favour the South-East. In 2003, a report from the Office of the Deputy Prime Minister concluded: 'At the moment our evidence suggests . . . that much government policy is, inadvertently, acting against the interests of the less prosperous regions' (ODPM, 2003, p. 22; see also Harding, Marvin and Robson, 2006). The government responds with more urgency and bigger money to the problems of congestion in the South-East (i.e., the demands of global city growth) than to the need for regeneration elsewhere. Unexamined assumptions reflect deeply embedded geographical imaginaries. Northern MP Louise Ellman reflects on major

science investments: 'The money invariably ends up in the
South East as that is seen as the "normal" place for science to
be' (*Planning*, 2001). In a variety of ways, often subtle, sometimes difficult to pin down, the concentrated spatial grammar of social and political power itself becomes part of the
process of the (re)production of regional inequality. Not
only are there relations between regions, but also those relations go way beyond the economic; they are utterly entangled with the national polity and society. And by no means
all of them are benign.

So one issue is power and democracy. But a further thread
runs on from this. The imaginary of the golden goose and all
those arguments that London must not in any way be challenged rest on a crucial assumption. This is that London has
'achieved' its present position through its own efforts. As the
hegemonic terminology has it: it is a 'successful' region. To
do anything to disturb London's trajectory would be to buck
market forces. This is, patently, only a part of the truth.
London's 'success' has had enormous help. Most obviously
the great turn-around in the 1980s towards the booming city
of today has been spurred above all by the classic measures of
'neoliberalism': privatisation, deregulation, and liberalisation
in general – processes that also contributed to the industrial
collapse of the North. In other words, the basis – the essential precondition – for the current growth of London is a
particular politico-economic conjuncture. The shift in dominance from public-sector professionals to private was geographically embodied in a shift towards the south-eastern
corner of the country. The dismantling of manufacturing by
finance (see Sampson, 2004, for the cases of ICI and Mar-
coni) was a victory of South over North. The privatisation of
previously nationalised industries could lead to relocation of
research and development towards the South-East (Hudson,
2006b). And so on. The elements of South-East bias in

national policy-making have also already been mentioned. But there is more. The fact that London is the unquestioned site of 'the national' brings to it a whole range of institutions. Many of these, precisely because they are 'national' (museums, sports centres, etc.) get at least an element of national funding, while institutions in the same sector but in other parts of the country will be dubbed 'regional' and will likely have to find their funding accordingly. It is simply because London is the (one and only) capital that it is assumed to be the 'natural' location for the national football stadium; it is only London that can be recognised as a possible bidder for the Olympics. The vast majority of national artistic and related institutions – museums, galleries, and so forth – are located in London. Moreover, the point about this is not only the location in itself, and what it does directly to reinforce London's growth – through employment, through the national funding that it attracts, through the tourism that it bolsters. It is also – and maybe in the long run rather more importantly – the inequality in ordinary lives that it fosters. This happens in a host of ways. Londoners have Tate Modern on their doorstep. The concentration of such institutions in the capital means that cultural 'news', as well as political and economic, is drawn to focus on what happens in that city. Children in London's schools have easy access to the Science Museum, while for those in, say, Barrow-in-Furness or Millom an equivalent school outing would cost a fortune in both time and money. In all these small but crucial ways the 'nation' is pulled apart by its very geography.

And finally, London gains advantage through a host of supposedly a-spatial national policies. The deregulation of mortgage finance fuelled a boom in owner-occupation, especially in the South-East, and as house prices there rocketed to levels four times higher than those in the North the South-East gained further, both from a disproportionate share of tax relief on mortgage interest and from the rise in its asset-base (Hamnett, 1989). Or again, Tony Blair goes to France to celebrate the new Airbus A380, an aeroplane so big that in its very design it encourages concentration on a few major airports – in the UK, London's Heathrow. Perhaps most important among these policies with no apparent geographical content, and talismanic in its economic effects, its symbolic meaning, and its relation once again to the resurgent elite, is policy on tax. London is the most unequal city in the country, and the South-East the most unequal region, and this is largely down to the presence there of the very wealthy. New Labour's deep-seated reluctance to address inequality *per se*, or the soar-away wealth of the richest, has led to a tax policy with no progressive element at the upper end. Inevitably such a tax regime, by default, pours money into the South-East. An effective tax on higher earners would not only in itself be more egalitarian; it would also in particular do something to address the inequality within London, and beyond that the inequality between the regions of the UK. This intricate connection between national levels of inequality on the one hand, and inequalities between regions on the other, is crucial. London is the site where such economic disparity is most acutely evident (and particular social groups and areas within London bear the brunt of this). But London and the South-East 'as a whole' 'benefits' (and is further divided) through the lack of high taxes on its highest earners. On the one hand, the 'social' and the 'spatial' are intimately intertwined. On the other hand, the difficulties and distortions produced by a territorially (regionally) organised politics start to become evident.

Again exactly in parallel to discourses at national level about 'welfare dependence' and such, there is the oft-repeated refrain that regions need to stand on their own feet (chapter 4). The Treasury/DTI report retails this view perfectly:

regions affected by large economic shocks do indeed 'require assistance', but, it says, 'Creating areas that are dependent on such assistance will, however, damage their long-term growth prospects and be detrimental to living standards in that area and the UK as a whole' (HM Treasury/DTI, 2001, p. 49). The background understanding is that a region must 'succeed' by market forces alone. The 'assistance' that London and the South-East have received is, of course, never called by that name, and it is not to downplay the evident energy of London to point this out. It is merely to point out that, while the regions of the North and West are instructed to adapt to market forces, London's unparalleled 'success' is by no means a product of those sanctified forces alone. Moreover the assumptions underlying this moral geography feed back to reinforce the wider neoliberal agenda. Thus Bosanquet, Cumming and Haldenby, in a pamphlet entitled 'Whitehall's last colonies: breaking the cycle of collectivisation in the UK regions', conclude that 'The main solution is a general one – to allow the growth of a stronger private economy through slowing down the growth of public spending and lowering taxation' (2006, p. 6).

Along with the question of interest rates, undoubtedly the most oft-cited conflictive relationship between London/the South-East and the regions and nations of the North and West revolves around flows of *labour*, and in particular highly qualified and graduate labour. Such flows of labour, like so many of these issues, are typical of global cities more generally. The London Chamber of Commerce and Industry touches on them in the second sentence of its document on the 'London deficit': 'The London economy is the largest and most successful regional economy in the UK', it says. 'It has often been suggested that its success has been to the detriment of other UK regions, drawing highly skilled people away from other areas. The reality is more complex'

(Hill, 2003, p. 1). Oxford Economic Forecasting (2004) dwell on the issue, and the foreword by the Corporation of London returns to it. It is a raw nerve.

London's growth trajectory, both over recent years and as planned for the future, implies an increase in demand for labour with degree-level qualifications. It is demand for this kind of labour that dominates the net increase in employment in the capital. London does not provide all of this itself and in consequence draws in professional people from abroad and from the rest of the country (see, for instance, Local Futures Group's research reported in Biles, 2001; Trades Union Congress, 2002; Woodward, 2002; Ward, 2004). It is a brain drain that has a double effect. Within London the dominance of demand for this kind of labour both makes it more difficult for Londoners without those qualifications to find work and, through the influx of higher-paid workers, increases the pressure on prices, thereby exacerbating inequality within the capital (chapter 2). From the regions and nations of the North and West it drains a stratum of the population that could be significant to their economic growth. There is, in other words, once again a link between the inequality within London and the economic difficulties of other regions. This has its vicious ironies. Gordon Brown lectures the regions that their regeneration should be led by the knowledge economy (Loney, 2001). Alan Johnson (minister for manufacturing), among many others, repeats the refrain that low skills are part of the regions' problems (Johnson, 2002). In other words, the regions are blamed for the losses they incur through feeding London's insatiable demand. And the government that berates them pursues 'national' policies that encourage the concentration on London and the South-East.

Now, this understanding of the migration of labour can be read from another angle. In this view, the move of graduates to London both makes best use of their skills, because of the greater opportunities and rewards available there, and promises those people the fuller development of their potential. This is an element in the more general 'escalator-region' argument, and it raises considerations that are not insignificant.[3] But even this, quite valid, argument itself raises issues of inequality: it assumes the inequality to begin with, it does not ask what happens to those who do not wish to migrate, and it ignores the vicious and virtuous circles the process engenders. Dorling and Thomas (2004) have mapped the geography of these occupational changes between 1991 and 2001. Thus, take the category 'managers'. Its internal composition has shifted, a decline in the number of small farmers in the rural regions being compensated for by a rise in the number of corporate managers. And these latter are both concentrated in London/the South-East and increasing fastest there. There have even been declines in some northern urban areas. For 'professionals' the picture is similar, although here there has been overall growth, and nowhere have there been declines. But once again London/the South-East, already in 1991 the region of strongest concentration, increased its numbers most rapidly. Nor is this due only to 'business/administrative' professionals: science and technology professionals, too, gathered even more into the sunbelt around London. (On all this see especially Dorling and Thomas, 2004, chapter 6.) Even *associate* professionals are becoming more concentrated around London although, as ever, they remain more evenly distributed than their more highly paid counterparts. The vicious and virtuous circles these shifting geographies engender selectively open up and close down the possibilities for the regional futures. Take, for instance, the recursive relationship between occupations and qualifications, and their geographies. Dorling and Thomas report 'quite remarkable geographical inequalities in qualifications attained across the Kingdom. . . . in most of the larger towns of the North of England, in South Wales, Central Scotland and almost all of Northern Ireland, up to a third of the population hold no educational qualifications. In most of the South less than a sixth of the population are in the same situation, despite the concentration of many retirement areas in the South' (ibid., p. 81).

This is to speak of numbers; of occupations and qualifications. But these quantities and categories represent also functions within the economy. Different economic phases, and different politico-economic strategies, are established through distinct geographies. The particular jobs implied by 'sectors of production', or maps of the distribution of occupations, imply also a geography of roles within the economy more widely: a spatial division of labour (Massey, [1984] 1995). And between the different functions, and thus their geographical locations, run the differentially empowered social relations of production. Corporate managers, so concentrated in London, represent strategic decision-making within companies; they are the ones who negotiate with government; have political and economic influence; take decisions that change the lives of those in other, subordinate, occupations (and, as we see, in other, subordinate, places). These geographies of occupations are also geographies of power and influence. The current political and economic trajectory looks set only to reinforce not only geographical inequality but, through the further reverberations of that very geographical unevenness, national levels of inequality too.

6

WHO OWES WHOM?

London, then, is a 'successful' 'global city' set within a national context of increasing inequality. Indeed its own growth would seem to be implicated in that increasing inequality. On the other hand, however, London itself is also riven with inequality, and encounters serious difficulties in reproducing itself socially and maintaining itself physically. The driving force behind both elements in this doubled geography of inequality has been the turn to neoliberalism and the concomitant rise to dominance of finance and associated sectors.

This is a geography that has drawn a range of political responses, among them one (a dominant one) that demonstrates to the full both the power of the 'new elites' and the significance to them of the mobilisation of their own geographical imaginaries. It is therefore worth particular attention. Moreover it also raises further questions about the construction of the identity of place, the territorialisation of political constituencies, and the ambiguities of political localism.

The issue is that of the 'London deficit' – the argument that London has been subsidising the rest of the country and can afford to do so no longer (or certainly not to quite the same extent). Once again, imaginaries of the golden goose are crucial.

A report for the London Chamber of Commerce and Industry (LCCI) and entitled *The London deficit – a business perspective* provides an example:

> The London economy is the largest and most successful regional economy in the UK. It has often been suggested that its success has been to the detriment of other UK regions, drawing highly skilled people away from other areas. The reality is more complex. As will be seen from this report, the UK's progressive taxation structure ensures that London contributes a greater proportion of total income raised from taxation in the UK than any other region. In short, London subsidises the rest of the UK, enabling the nation as a whole to benefit from the capital's success.
>
> . . .
>
> The London Chamber promotes and campaigns for the needs of London's businesses and the London Deficit issue is central to this. Whilst the Chamber accepts that, as the economic epicentre of the UK, London should provide the lion's share of revenue, it should not subsidise the UK to the extent that it currently does.
>
> The burden placed on the capital's economy is an inequitable one. Consequently the Chamber seeks to campaign for a more egalitarian position. (Hill, 2003, pp. 1 and 4)

There are some immediate points to note about the way in which the notion of deficit in general constructs the imaginative geography of the economy.

Most significantly, it focuses tightly on very specific monetary flows. These are the only inter-regional relations of which account is taken. The London deficit is defined as the difference between government revenue raised from London, on the one hand, and the level of government public spending within London, on the other hand. The capital pays more in tax than it receives in public spending, the amount usually being estimated to be between £15 and £20 billion per year. The calculations are not simple; and there are various bases on which it can be made (residence or workplace, for example). A lot of energy has been invested in debate over the minutiae of the construction of these figures. Meanwhile, the larger geography is successfully obscured.

For this narrow focus ignores all the other relations that connect these regions. This can be seen even in the tortured detail of the technical debate itself. Thus, public expenditure on the national Civil Service is usually included in the figures for London, for the bulk of the Civil Service is indeed in the capital. The figure for public expenditure in London is thereby raised. It is also the case that these expenditures pay for a Civil Service that operates over the country as a whole. The argument of those who wish to press the claim for reducing London's deficit is that this is a service provided *by* London *for* the nation as a whole. This role of being the capital city is thereby constructed as a *burden*, and the public money apparently being invested in London should not count as being *for* London. This is an imaginary that shares much with that of Oxford Economic Forecasting (see chapter 4). But it ignores the more complex geographies within which these monetary flows are embedded. For the presence of such a weight of Civil Service jobs and functions within London contributes significantly, not only to London's economic growth and vitality (and through both direct and indirect effects), but also to the setting in place of a goodly part of that spatial grammar of power and politics which is not only at the seat of London's unequal relation with the rest of the country but also helps mould the very way in which national policy, including national economic policy, is drawn up. It obscures the benefits that accrue from simply being the national capital. On page 18 of the LCCI document a new section is headed 'The cost of being the Capital City'. To separate out a few monetary flows from this complex geometry of power relations is seriously to miss the bigger picture. It is also to imagine regions as already constituted territories between which flows may pass. In fact such flows are part and parcel of wider relations through which the regions are continually constituted.

The issues, however, go deeper than this. The *reason* that this argument is made about the London deficit and the need to reduce it is that there are indeed genuine problems within London that do require expenditure. This is the basis of the LCCI's case. Immediately after the opening statement, just cited, the document describes the problems that, in spite of its great wealth, beset the national capital. Figures are presented to demonstrate the serious presence within the city of wards with the highest levels of deprivation; unemployment is documented; and the inequality spelled out: 'There is great disparity within the capital, where the wealthiest 20% have incomes more than seven times higher than those of the bottom 20%. In comparison, the average differential elsewhere in the UK is less than five times, suggesting greater earnings in London but also greater deprivation' (Hill, 2003, p. 7). None of this is in dispute (see chapter 2). What is missing is any recognition of the mechanisms behind the *production* of this inequality.

The document goes further in its analysis of 'London's plight' (Hill, 2003, p.17) and argues that there is in fact a 'South/North Divide': that the old pattern of regional inequality has been reversed. The case rests on regional comparisons in which costs are set against incomes. In other words, London people and London businesses may earn more, but this is more than countered by the higher costs of living in the capital. Again this is not in doubt (see chapter 2). Thus: 'in a survey measuring local income against regional living costs in England and Wales, eight of the ten wealthiest areas were located within the North of England. Whilst only one London borough (Kensington and Chelsea) featured within the top ten, 13 were graded in the bottom 20 in terms of residents' purchasing power (Barclays Private Clients, May 2003)' (ibid., p. 11). Moreover, and again in agreement with the arguments of chapter 2, the most significant contributor to the high level of costs that produces this situation is said to be property prices. This is recognised to affect both public and private sectors, both social reproduction and economic growth. On the one hand 'London's public provision will suffer, with those workers delivering essential services such as health, education, transport and teaching unable to afford a home in the capital' (ibid.). On the other hand, there is the problem of commercial property prices. The main focus here is on office prices, and the document argues: 'if the premise that "the office has become London's factory floor" is true, there are serious implications for London's business in terms of the relative costs of managing a business in the capital compared to outside it' (ibid., p. 12). While the aim here is to construct a case for London as a whole, the divisions – even within capital – are occasionally evident. Thus a footnote to the last quotation acknowledges the LCCI's disparate membership: 'The London Chamber does not however support this premise [that London is all offices], promoting a balanced economy comprised of a strong service (including financial) *and* production sector' (ibid., emphasis in original). Moreover, as well as discomfiture between services and production, there is also the question of property capital itself, which owns this expensive property and benefits from that ownership. In the same paragraph from which the earlier quotation was taken there is reference to 'the ever present *threat* of a property market slowdown' (emphasis added). Indeed, some of London's deficit with the rest of the country is derived from the redistribution of rates based on ownership of non-domestic property – the National Non-Domestic Rate (NNDR): 'As a result of the high property valuation in London, NNDR is paid at a disproportionate rate by London's businesses' (ibid., p. 23). Unaddressed, but bubbling under the surface, is a more complex understanding of London's economy as an entanglement of interacting, and sometimes conflicting, trajectories. The fact that these are jarring interests is not recognised explicitly. Conflicting interests are buried under a language of simple diversity – 'a strong service (including financial) *and* production sector'. Nor is there any serious address to the way in which these burdens upon London are a direct result of the fact, and especially the nature, of its own inexorable expansion. The possibility of refraining from encouraging such expansion is not mentioned. Still less is there any serious exploration of the implications of the unequal way in which these high costs fall, their ongoing entanglement precisely with the stark inequalities within London from which the report builds its case.

What is at issue here is the construction of the identity of place. A divided London must be sutured into the singularity of a 'we'. This is a necessary task in the sense that the throwntogetherness of place (Massey, 2005) requires that spatial contiguity is addressed, that negotiations take place, that collective decisions are reached. There are, however, different ways (different *politically*) in which the identity of place can be constructed. In the case of the London deficit the identity presented is not a 'we' of multiple and conflicting interests, of clashing trajectories. It is not an identity, such as that evoked in the days after the bombing, of the recognition of difference and of the need to negotiate – an identity always in the process of construction. Nor is it an identity achieved through the grounded negotiation of a public. It is in no sense democratically arrived at, but rather assumed by a hegemonic voice. This is the assertion of a London constructed as a spatial unity through opposition to an outside constructed as an 'external enemy'. In truth, 'enemy' is here the wrong term; the attitude is rather one of condescension towards the rest of the UK. It is to be regretted, but we are very much afraid that we can no longer afford to subsidise you so much. We know you have problems but so do we; we know indeed that we are way ahead in the economic game, but at the moment we simply cannot afford to help out as much as we have been doing. This is not a competitive enemy (as in the competition with other global cities). It is not even a hostility; if anything it is patronising. Indeed it exactly exhibits those attitudes so deeply embedded in the national geography of power. It is a geographical imagination of a particular form of territorialisation, and one that is worked to political effect. 'If *London* is expected to finance the UK economy, *London's* own economy must be supported to fund this' (Hill, 2003, p. 34, emphasis added). The agents here are territories, regions. It is 'London' that pays too much tax, and that gets too little back.

However, the reason 'London' pays so much tax is that there are some within it (both individuals and businesses) that are stupendously rich. (One feels the reference to the existence of a 'progressive taxation structure' must be ironic, given how much London and the South-East gain from the lack of such a structure.) And London is also the most unequal region in the country; the deprivation within the capital is not in doubt. Instead, then, of a London united across its inequalities in order to make claims upon the North (Liverpool? Wallsend?) what might be more effective is to examine the possibilities of *redistribution within London itself*. It is here that wealth and poverty sit most glaringly side by side. 'The likes of Oldham, looking south, might legitimately point to the longstanding juxtaposition of "City" and "East End" and tell London to put its own house in order' (Amin, Massey and Thrift, 2003, p. 21). The LCCI document, as we have seen, makes its pitch on the basis of claims against inequity . . . it 'seeks to campaign for a more egalitarian position' (Hill, 2003, p. 4). It is a highly specific reading of egalitarianism and one that (even while its case is based on them) occludes the most glaring inequalities of all, those within London itself. In effect these claims for a diminution of the London deficit amount to a call for the rest of the country to pitch in to ensure the profits of 'London-global-city' by paying for the costs of congestion and inequality that its own roaring expansion has played such a large part in producing.

In 2003 I took the opportunity of a generous invitation to make some of these arguments at a conference hosted by the Chamber of Commerce itself. It was a conference specifically concerned with the London deficit, and the invitation acknowledged in a friendly way that I might present a contrary view. The meeting was held amid some of the more sumptuous displays of wealth in the country. It was difficult not to see the irony of the setting for a debate on 'London's plight'. I mentioned this in my talk, and made the arguments. Those comfortably in power can afford not to get too annoyed. They can smile, even be amused, by points of view they know will get nowhere, at least in the establishment circles where they know their influence holds sway. And so it was on this occasion. The arguments were heard out, but the meeting did not seriously engage the debate.

Far more serious than this little attempt to speak truth to power has been the ongoing organisation of the Living Wage campaign. This was first established in 2001 by TELCO (The East London Communities Organisation), an alliance of civil society institutions including faith groups, trades unions and community organisations (Wills, 2004). Its initiating aim was to pick up on the principles of the Living Wage movement in the USA, and its targets include both public- and private-sector employers. One of the first places to which TELCO took the campaign was

> the prestigious office complex at Canary Wharf. Since it was built in [the] 1980s with massive public subsidy, Canary Wharf has come to symbolise the social and economic inequality of London's East End. Whereas financial and media professionals work there during the day, often for huge salaries and lavish benefits, low-paid cleaners labour during the night, for as little as £4.50 an hour without any additional perks . . . To focus the campaign, TELCO targeted the new global headquarters of HSBC that opened in mid-2002, demanding that the tower be cleaned by those earning a living rather than a minimum wage. (Ibid., p. 278).

The supporting actions have been varied and imaginative (see table 6.1). The initial responses were mixed but telling: 'Living Wages have been backed by private employers such as the audio retail chain Richer Sounds, but TELCO has had less success with city banks. A campaign to force HSBC to insist its contract staff receive the £6 minimum was rebuffed' (Walker, 2003). An attempt to hold a demonstration for living wages from these sectors of London's economy, made while London was hosting the European Social Forum in October 2004, was likewise 'rebuffed', the demonstrators at Canary Wharf not even being allowed to assemble, on the grounds that this land was (now) privately owned. The response that one should not 'interfere in the market to pay cleaners more' (see table 6.1) from an organisation which within less than twelve months would appoint a director on £35 million has a telling ring about the operation of 'market forces', as does the geography of the trajectories that were meeting up here in Canary Wharf: the director was from the USA, the cleaners from a wide range of countries in the global South and in Eastern Europe. In that light the LCCI's description of the city ('One of the world's leading centres for global business, it is however a capital with a vast disparity of wealth'; Hill, 2003, p. 5), its stated commitment to 'a more egalitarian position' (ibid., p. 4) and its recognition that the 'vast disparity' is making London difficult to run is lent more than a touch of irony. Certainly it did not seem to lead all its members to recognise their complicity in that inequality or to address the possibility of themselves con-tributing to its alleviation. Instead they demand money back from the rest of the country. The living wage campaigners, however, were not to be put off, and slowly significant gains have been won, including through agreements with both Bar-clays and HSBC in the finance sector (see table 6.1).

In a televised debate in June 2001, Ken Livingstone him-self became embroiled in the contradictions of the claims of a London deficit. Livingstone, as mayor of the whole of London, is also, given the territorial framework in which he is operating, pushed to argue a London-wide position. The London government too, in other words, has argued that the poverty in the city means it needs to claim back more of the national tax-take. The tone, however, is noticeably dif-ferent from in the private sector; and it has modulated over

Table 6.1   A chronology of the London Living Wage campaign

Date	Events
November 2000	Retreat of TELCO leaders discussed common concerns and focused on poverty, privatisation and the challenges on people's time. Discussed the experience of the living wage campaigns in the USA and floated the idea.
January 2001	UNISON commissioned the Family Budget Unit to determine the living wage rate for London (£6.70 at the time of writing).
September 2001	Published *Mapping low pay in East London*, documenting the extent to which low-paid workers are in the gap between the minimum and living wage, with very poor conditions of work. Launched at a conference held at Queen Mary, University of London.
November 2001	Public assembly of up to 1000 held at York Hall, Bethnal Green, attended by local MPs, officials from HSBC and John Monks (general secretary of the TUC), who restated the case for a living wage.
December 2001	Occupied a branch of HSBC in Oxford Street to protest at the low pay and poor conditions of cleaners at the Canary Wharf site.
April 2002	Parliamentary hearing for the living wage campaign, held at the House of Commons with invited MPs and a guest speaker from Baltimore.
May 2002	At least 40 ISS workers, along with representatives from local mosques, churches, colleges and the media, attended a meeting of the NHS Trust board at the Homerton Hospital, made a presentation and handed over a petition with 600 signatures from hospital staff.
May 2002	Attended AGM of HSBC to demand a meeting with the chairman, Sir John Bond, to discuss contracting arrangements at Canary Wharf. TELCO activists had bought shares and were able to interrupt the meeting, asking for action.
June 2002	Meeting held with Sir John Bond, who rejected the idea of interfering in the market in order to pay cleaners more.

Date	Events
July 2002	UNISON/TELCO submitted a claim for improved conditions for staff working for the contractors ISS Mediclean and Medirest at five East London hospitals.
Autumn 2002	Living wage campaign supported by Billy Bragg's tour, raising the profile and finance from gigs across the country.
November 2002	Held living wage march involving local schools along the Mile End Road before large public assembly at the People's Palace, Queen Mary, University of London. Led by black Pentecostal churches from the area. Focused on the NHS and developing a relationship with the mayor of London.
March 2003	Held academic conference at the London School of Economics to make the case for a living wage, drawing on US experience.
April 2003	Public rally of 200 contract workers involved in the East London claims held in Stratford, addressed by the general secretary of UNISON and workplace activists, building for industrial action in pursuit of the claim.
May 2003	Public assembly of about 500 held in Stratford, to confirm support for a strike of contract staff in the hospitals. New aspect of the campaign also developed to work on better ethical standards in UK contracting practice.
May 2003	Demonstration outside and inside HSBC's AGM, again putting pressure on the bank over contracting arrangements at Canary Wharf. Generated a lot of media attention, and the pressure coincided with the appointment of a new director from the USA, who was to be paid £35 million.
June 2003	Improved offer accepted by contract staff at the Homerton and Mile End/St Clements hospitals, to secure immediate improvements and parity with NHS conditions by 2006. Strike and large demonstration held to try and increase the wage level and improve the offer at Whipps Cross Hospital. Later offer made to Royal London contract staff.

Date	Events
September 2003	Discussion of plans to launch an initiative for socially responsible contracting with cleaning clients, the industry, unions and ethical investors, held at the DTI.
November 2003	Launch of *Socially responsible contracting* document at Portcullis House, Westminster. Chaired by John Cruddas MP, attended by Barclays, major banks, cleaning companies, ethical investors and MPs. Estimated to cost 30 per cent more for contracts that meet standards.
February 2004	Barclays Bank agrees to new terms and conditions for contract cleaners employed at new HQ in Canary Wharf. Pay to rise to £6 an hour, fifteen days paid sick and eight extra days holiday. Access to the pension scheme also included. Reported by BBC Newsnight; good media coverage.
	Around this time, the Transport and General Workers' Union deployed two full-time union organisers to work at Canary Wharf.
May 2004	Mayoral Accountability Assembly, Westminster Central Hall, calls on all candidates to support the establishment of a Living Wage Unit to publish an annual living wage for the capital. Ken Livingstone, the successful candidate, agrees. Barclays praised publicly at the meeting.
May 2004	OCS Cleaning announce a new package for staff at HSBC Canary Wharf (produced in negotiation with HSBC). Wage increase of 11 per cent to £6.10 an hour, eight extra days holiday and ten days sick pay. Changing shift patterns to reduce night working and those doing nights to be paid 30 per cent increase. Encouragement for training and career development. Invitation to join OCS pension scheme. Agreement made two weeks before HSBC AGM.
November 2004	COMPACT signed between London Citizens and the London Olympic Committee to endorse an ethical Olympics. Agreement includes paying a living wage to contractors working on the Olympic construction and those employed during the events.

Date	Events
March 2005	The Living Wage Unit (GLA) publishes *A fairer London* and announces poverty threshold wage of £5.80 an hour and a living wage of £6.70 (including benefits and tax credits) for the capital. One in five workers found to be on less than the living wage. Mayor agrees to roll out the living wage to the GLA/Transport for London and to include it as a criterion in grant-giving process.
May 2005	General election used as an opportunity to hold accountability assemblies, with the main candidates in each constituency asking them to support a living wage along with other TELCO/London Citizens demands.
July 2005	London Citizens – Queen Mary, University of London Summer Academy project to research the pay and conditions of workers in other low-paying sectors of the economy: care, hospitality, transport cleaning, office cleaning. *Making the city work: low paid employment in London* launched at South London Citizens Assembly.
August 2005	IPPR became the first third-sector organisation to address the living wage for cleaners, and employ ethical cleaning company.
September 2005	T&G steps up its organising campaign in Canary Wharf. Living wage breakthrough at Deutsche Bank in the City. Followed by Morgan Stanley, Lehman Brothers and Citigroup at Canary Wharf, KPMG, PricewaterhouseCoopers, RBS (mainly in 2006) in the City. Starts to move onto the underground network.
October 2005	Living wage organising effort started at Queen Mary, University of London, with rally held to win improved conditions for cleaners at KGB Cleaning Support Services.
February 2006	Parity with NHS terms and conditions secured for contracted cleaners at Homerton and Whipps Cross hospitals (the latter only after more industrial action).

Date	Events
April 2006	Research published into impact of living wages paid to domestics at the Royal London Hospital, as staff brought in-house via a PFI deal in summer 2005.
April 2006	Queen Mary declared to be the first living-wage campus in the UK and campaign taken up at the London School of Economics.
May 2006	May day mass to honour migrant workers attracts up to 2000 people to Westminster Cathedral. Homily given by Cardinal Cormac Murphy O'Connor with call for regularisation, taken up by London Citizens rally outside. Founding of London Citizens Workers' Association (LCWA).
May 2006	TELCO and London First organise conference for the industry, to discuss efforts to raise standards: *Keeping London clean and well*.
May 2006	GLA publishes *A fairer London*, updating the living wage figure to £7.05.
July 2006	London Citizens and UNITE-HERE work together on hotel workers' campaign to take the living wage to London's hotel sector by targeting the Hilton group and Kensington Close. Mass action at Hilton Metropole to demand meeting with Howard Friedman, the president of Hilton International in the UK. Ongoing recruitment of workers to London Citizens Workers' Association from LSE, Tate Modern and Hilton group.
September 2006	Draft Olympic delivery procurement policy found to have dropped commitment to a living wage. Rally held to target David Higgins, chief executive of Olympics Delivery Authority (ODA), for a meeting.
November 2006	Meetings held with managers from Hilton Group about employment conditions.
November 2006	TELCO tenth Anniversary assembly; Queen Mary announce plan to bring back cleaning in-house and exceed living wage demands; ODA promise to meet and discuss the living wage. Sir John Bond is honoured on stage for his willingness to work with TELCO to make improvements at HSBC, despite his initial hostility.

Date	Events
November 2006	T&G step up efforts for a zonal agreement for cleaning standards across Canary Wharf and the City; sign up ISS, OCS and Lancaster.

Source: http://www.geog.qmul.ac.uk/livingwage (regular updates are made).

the years to adjust to the difficult politics of inequality here and to reflect its own political stance. It is clear that for a radical mayor this is a minefield. Nonetheless, he goes out to bat for London. It is the imperative of inter-place competition. The television programme in 2001 was entitled 'Great London rip-off', and in a commentary the following day the London *Evening Standard* began:

> Ken Livingstone today stands accused of attempting to create a divided Britain after his new campaign for greater government funding for London received a hostile response from political leaders in other regions.
>
> The Mayor put his case to the court of nationwide public opinion in a televised debate last night, during which he argued that the whole of the country will suffer unless big injections of cash are put into London's creaking infrastructure. (Freeman, 2001)

Here, it should be noted, the case for London is being made, not on the grounds of the poverty within it, but rather on the basis that investment in London is to the benefit of the country as a whole. It is the 'golden goose' argument. Nonetheless, the TV debate for a while descended into a poverty competition:

> much of the debate ended up being taken up with conflicting 'sob stories' over who was the worst off.

The North East claimed to be the poorest part of Britain, while Humberside claimed to get the least subsidy of all.

A London GP's tales of severe child health problems in poverty-stricken East End boroughs were met with Scottish claims to having record infant mortality rates. (Ibid.)

In other words, framing the debate in this way, setting region against region, also effectively divided the working class. The representatives of the poor squabbled among themselves. In spite of having himself established the debate in these terms, this was not an implication that Livingstone could support. 'I am not interested', he said, 'in taking money away from poor people in the North or anywhere else.' To which Alex Salmond, of the Scottish National Party, responded: 'Why don't the opulent areas of London pay a little more council tax than they do now to subsidise the poor ones?' (ibid.).[1] This, in general terms, is the crucial question. And Livingstone replied: 'I would like that, but I haven't got the freedom to do so' (ibid.).

It is difficult not to feel sympathy for this position. London is indeed in a spatial trap. First of all, in part Livingstone is quite correct. New Labour's refusal, for instance, to raise the level of taxes on very high incomes, their condoning of fat-cattery and their lack of commitment to reducing inequality in the country as a whole certainly contributes to the problems in London, and Ken has no control over that. Nor can he raise money from commercial interests through the business rate as the GLC did in the 1980s.[2] Second, however, 'London' has a powerful voice which could be used to argue this case, as it was used in a whole host of issues by the GLC in the 1980s. The effects of nationally increasing levels of inequality bite particularly fiercely in the capital. It might also be argued, and is by many, that the GLA should have more tax-raising powers and thus a greater basis for redistribution within London. It is also the case, third, that many of the GLA's policies under Livingstone are devised explicitly with precisely that redistribution within London in mind, and the GLA has indeed adopted a Living Wage Strategy for its own employees (negotiations continue about its further extension) and now publishes an annual report, *A fairer London*, including a calculation of the living wage for London (GLA, 2005c and 2006) which, although it cannot simply be enforced across the local economy, can provide a benchmark and a bargaining tool. A whole range of interventions is explicitly geared towards redistribution within London itself. However, fourthly, redistribution does not necessarily have to be *post hoc* (first economic growth and then redistribution). The *form* of economic growth can itself influence distribution in the first place. And the current form is exacerbating inequality. In this sense too there is some room for local manoeuvre.

There is no doubt that the framing of this deficit debate in territorial terms, as a form of competitive localism, has been problematical for a progressive politics. It can immobilise the trades unions as regions are pitted against each other and it deflects from the more general politics around inequality that should really be at the centre of debate. Yet across the political spectrum, including the political and grass-roots left, there is a renewed enthusiasm for 'localism'. This is a form of territorialisation that is seen as intrinsically more democratic; it certainly provides an arena for grass-roots organising; sometimes it is held out as the sphere of lived particularity that must be defended against the global outside. There are many cautions that may be raised against this new enthusiasm, including that it is not necessarily more democratic (see Amin, Massey and Thrift, 2003), but the case of the London deficit points to two specific difficulties.

First there is the question of how 'the local', the identity of place, is to be constructed. This concerns both the manner of arriving at a hegemonic position – whether it be by political debate or by simple assertion on the part of the powerful – and the nature of the identity that is constructed – closed and defensive or interactive and recognised as continually in process. Second there are the problems of spatial fetishism implicit in support for the local *qua* local (see Massey, 2005) – in particular here the imaginary of 'the local' as always better ('local people'), more authentic, with more rights to have their say than others 'outside' (i.e., wider, more general, interests). In the case of the London deficit the localism is taken up by the relatively advantaged locality. The powerful can play at localism too. Not only are the difficulties of redistribution exacerbated, and political constituencies that might better be working together torn apart (which can happen with any form of localism), but the already powerful (richer, more influential, . . .) localities stand to gain even more. The already strong will become stronger. Localism cannot be a general principle unless it addresses also the localism of the already advantaged. Support for the local place as such does not bring with it the implication of any particular politics.

7

REWORKING THE GEOGRAPHIES
OF ALLEGIANCE

The argument over the London deficit assumes that the problems of poverty within London and the problems of regional inequality are *competing* political priorities. It is an implicit assumption that runs as a current of shared understanding through many other lines of policy debate. Yet in fact the two are intimately linked. The depth of the poverty within London, and the difficulties of the social and infrastructural reproduction of the capital, are in part a result of the inequality within the city (chapter 2). It is an inequality that is fed both by national policies (including broad political stances) and by the particular nature of London's reinvention. And elements of this reinvention are bound up with the reproduction and exacerbation of the 'North–South divide'. In other words, the *nature* of London's growth is part of the dynamic behind the current mode of reproduction of *regional* inequality. Behind everything else lies the rise of London as 'global' neoliberal city, and the class victory that that represents. *Both* the poor within the metropolis *and* the majorities in the regions of the North and West could benefit from an alliance which challenged the spiralling wealth, and the social and political dominance, of a London-based few.

One of the difficulties of constructing such an alliance, and such a challenge, derives from the plausibility of the geographical stories that are told in order to legitimise the (newly achieved) status quo. These dominant stories frame the debate, fixing the imaginary of both city and nation. London is characterised as a global city (yet that is only one aspect, and not the most significant, of its identity, even its economic identity). Both the City and the whole burgeoning institutional infrastructure of neoliberalism within London, and London within the nation, are characterised as golden geese, their benefits flowing out to 'the rest' across the smooth spatial surface of the economic imaginary of trickle-down (yet that occludes both the agency of others and the possibility that not all these effects are benign). Cities and regions are imagined as singular agents (thus ignoring the conflicts within). They are (in some cases) instructed to stand on their own two feet (thus ignoring the power-filled relational space in which each is embedded).

The left, too, has sometimes been caught up in this entangled situation, in both its implacable materiality and its discursive framing. Such imaginaries, on which the peak organisations of capital in the capital have persistently insisted, make it difficult to raise questions about conflicting interests within London and its region or about the potentially negative implications in the rest of the UK of this particular reinvention of the capital. It is indeed not easy to frame a response to such imaginaries, especially given the constraints of inherited structures, allegiances and conventions. National trades unions have on occasions found themselves immobilised as a result of the apparently conflicting interests of their constituent regions. The radical instincts of

some London governmental representatives have been cor-
ralled, and thereby constrained, as the result of the pressures
and demands of a territorially organised politics. Some progressive local groups in London have ended up pleading special cases for the capital. It has, in other words, been quite
difficult for some elements of the left, within London especially, to avoid becoming involved in a political geography
that pitches North against South, and vice versa.

This is by no means always explicit, or intended. The
example of my own trades union (then the Association of
University Teachers) provides a case in point. One of the
'special measures' (see later) by which the spiralling costs of
London's growth have historically been warded off has been
'London weighting'. Public-sector wages in the United
Kingdom have historically been based on the principle of
nationally equal earnings for each job. It has, arguably, been
one of those threads in the dense tissue of understandings
and practices that has underpinned the very constitution of
that element of the notion of 'the public' that relates to the
public sector. As a nurse, or a teacher, or a social worker, you
get paid the same rate for the job wherever you are in the
country. There has, however, been one major breach in this
geographical solidarity. Public-sector wages are not high.
Certainly they are not high in relation to regions with an
above-average cost of living. In response to this, London's
particularly high cost of living was used to lever open the
system, and workers employed in the capital came to be
awarded a special supplement: the London weighting. In
2002–3 the London area of the Association of University
Teachers came out on strike for an increase in their London
weighting. They were not the only public-sector union to
address this issue. Surviving on public-sector wages in
London is difficult (though far easier, it has to be said, for
those on the wages of a university teacher than for most in

the public sector, and for the very many low-wage workers in the private sector too). Public-sector workers were caught between the spiralling costs of living in the new London and the disparagement of the public sector that was another side of neoliberalism. So times were hard.

Nonetheless the decision to come out on strike for an increase in London weighting raises some important political questions. Most obviously, this was not a strike for higher *national* public-sector wages; it was specifically a claim for increasing wage differentials with workers in other parts of the country. It was in that simple sense divisive. Moreover, this claim was made at a time when the New Labour national government wanted, more generally, to break up the old system of national wage rates by introducing regional wage bargaining in order to drive down wages overall. This was being fiercely resisted by trades unions, including those – such as the AUT – which were at the same time demanding regional differentials for London. For the trades unions this was a classic spatial trap. And in that sense the London claim was a trades-union localist politics that did not coherently set itself within the geographically wider issues. The effect, indeed, would be to contribute to drawing the wage/price structure of the London economy yet further away from that of the rest of the country. And beyond the purely economic calculus, it would contribute to the further fracturing of a national sense of the public. Finally, and most dispiritingly of all, it would anyway not in the end solve anything. Increasing the London weighting was just another way of responding to the concentration of national growth in London, but as such it was also a way of shoring up the possibility of that concentration continuing.

Insofar as the left responded in this manner it became caught up in, and divided by, a territorial politics and a territorialised imagination of the national geography. In the absence of any address to the wider structural issues this was difficult to avoid. These issues are not easy for the left. The Living Wage campaign's successes in London (chapter 6) have led to the establishment of a similar campaign in Birmingham. And the idea is being suggested more widely on the left as part of a national strategy (Compass, 2006). This poses real questions. What will happen if different local areas agree different wages? Will that not work against the defence of a national public and solidarity? Will it not help institutionalise regional inequality? These are real and difficult issues in the geography of left politics. They are further complicated in this case by an additional spatial complexity: that the aim of sustaining a national public imaginary may anyway be less shared among the poor and marginalised in London because such a high proportion are migrants from abroad, and often temporary migrants at that.

One response to the difficulty of arriving at geographically differentiated living wages (and that differentiates this from the AUT case) is that these campaigns, although they are about money wages, are also about more than that. They provide a way of contacting migrant workers heretofore so often beyond the reach of trades unions. But above all they are about dignity and *respect*. That is indeed the difference between a living wage and a minimum wage (Wills, 2004). The lack of human dignity and the simple lack of visibility of these workers is a constant refrain. They work at night when the offices are otherwise empty, their invisibility exacerbated by the fact that this work is usually contracted out (Allen and Henry, 1997; Wills, 2004). In London, their poverty and invisibility represent a precise inversion of the deliberately conspicuous greed of consumption at the other end of the scale. The Compass programme also stresses this aspect: 'We believe that wages and benefits should be high enough to be compatible with human dignity. . . . we advocate a living wage. . . . The inequality gap should not be so large as to prevent recognition, or to fracture the bonds of common citizenship' (2006, p. 24). Nonetheless, the issues of geography remain, and they will have to be grappled with as the campaigns spread. Compass argues that 'A standard living wage should be introduced across the country' (ibid., p. 41). If it is, it must be high enough to be a living wage, with dignity and respect, in London too.

The challenge is for local struggles to avoid local*ism*. There are many positive aspects to a locally oriented political focus, but some of its dangers are equally obvious: a failure to take on board the opening this potentially offers to a counter-localism by the powerful (see chapter 6); a sinking into a romance of the local; the tendency to construct an identity through external enemies that in fact include potential allies; an assumption that 'bottom-up' politics is *ipso facto* progressive. But there are bigger questions too that can be posed to this conceptualisation of local struggle. First, the common association, so frequently made, between the local place and the authenticity of daily life is not only questionable in a world of flows (see chapter 8), but it can also set up the non-local as not embedded, as somehow more abstract; it can make it more difficult to think *beyond* the local. Second, this thinking of the local as uniquely embedded can encourage a certain closure of identity, an understanding of identity as pre-formed before engagement with the world beyond. Yet every local place and local struggle will be also a weaving together of wider influences, both those that have gone into producing the local place (and its identity) and those wider political conversations, contacts, understandings that in fact inform the very nature of the local political struggle itself (Featherstone, 2001). This is not an argument, in any way, against locally based struggles; nor is it a claim that all of the locally based struggles in London are of this nature. It is nonetheless the case that certain ways of framing the nature of 'the local' can make it more difficult to open up to a politics beyond.

In an age of increasing, and increasingly extended, geographical flows and connections (and indeed where identity has been reimagined relationally), this understanding of space as constitutively territorialised becomes more and more problematical. A more relational imagination of space, and thus of local place, could underpin a more networked, configurational, local politics.

Certainly it is necessary to go beyond that framing of the question of national geography in terms of North *versus* South, or in terms of London *versus* the Rest of the UK. The North–South divide is more than ever a distillation of *national* levels of inequality. The biggest interests of ordinary people, in both London and 'the regions', are in common. Neither regional inequality nor poverty within London/the South-East will be seriously addressed without a shift in the national model of economic growth (which is also the basis of London's reinvention) and an attack on national levels of inequality.

It has been argued that the problem of poverty in the North is one that requires regeneration, while the problem of poverty in London is one that requires redistribution. Given its present geography, a reduction in *national* inequality would be some contribution to addressing both. Around this kind of issue many 'local' struggles in London/the South-East and North and West could unite. There is a range of other possibilities. The essentially London-based campaign against the Olympics could have added to its collection of objections the fact that staging such an event would also further feed inter-regional inequality. National campaigns against the deregulation of the labour market could add to their armoury of arguments a regional dimension. Perhaps above all, local struggles of the poor in London and activists against regional inequality could join in questioning the centrality and power of the City constellation not only in the national economy but also in the national geography. They could likewise get together to point to that gross national geographical irrationality of paying at the same time the costs of regeneration (in the North and West) and the costs of congestion (in the South and East).

These would be a politics of place, yes, but one that recognised a commonality of interests in spite of the very different geographical positioning within the wider geographies. In this more configurational politics, 'local action' means action taken at a node within the multiplicity of trajectories that is the spatial, but with explicit recognition both of the construction of that node (i.e., that local place – it may itself be up for question; it may itself change) and of the wider ramifications (the linkages and implications of this politics beyond the place itself). A territorially grounded politics that is responsive to a relational space.

My argument here, then, while strongly motivated by a need to respond to the inequalities between regions, is absolutely *not* an anti-London one[1] – though it is in part to disaggregate that notion of 'London'. A more egalitarian and liveable London might be easier to achieve – indeed might entail its setting – within a regionally more egalitarian country.

*   *   *

It is not the purpose of this book to spell out policies to attain this regionally more egalitarian country. This has been done elsewhere (Amin, Massey and Thrift, 2003; Adams, Robinson and Vigor, 2003). The aim, rather, is to argue for a different way of approaching the question in the first place. New Labour has set the terms of the current debate: we mustn't interrupt London's 'success'; regions must compete against each other; we must not go back to 'old' (i.e., national-level) regional policy; etc. But it is these very terms that are inadequate to the task.

The spatial concentration of economy and society within one corner of the UK, and especially within England, is a centuries-long fact. But it has in recent decades been increasing in intensity. The nature of the relations that hold this geography together has shifted and become more tightly focused. Increasingly the country is being drawn into a vortex, centred on finance and the burgeoning business services that form the cultural infrastructure of neoliberalism together with a real-estate sector that both oils and benefits from their growth. This is both a reworking and a reinforcement of aspects of a longstanding national spatial division of labour and, again, one clearly tied in to (both dependent on and one of the nodes of creation of) the international economy. It is also a geography integral to the promotion of a very particular form of economy and society (Allen, Massey and Cochrane, 1998). And, as it stands at present, the broad sweep of policy is precisely to go along with, indeed to reinforce, that model.

However, within the power-geometries that construct this spatial division of labour, London and the other English regions, and Scotland and Wales, occupy very different positions. While each region is a porous intersection of a multiplicity of trajectories, and while all the regions are complexly interconnected, each is distinct in terms of its positioning in relation to these wider connections. They are differentially vulnerable to the effects of globalisation, they have different degrees of potential leverage, they are quite distinct in the degree to which, and the manner in which, they are the locations of the production of globalisation. The range of available practical policy options is very different between regions.

In relation to the long history of left debate over regional regeneration it is correct, when considering the regions of North and West England, to insist on 'not blaming the victim'. (And yet this is what New Labour continues to do.) The balance of the power-geometries within which these regions are set means that they have much less autonomous room for manoeuvre within, or leverage over, wider global forces. London, however, is in a very different position. And it is for this reason that Part III (chapters 8, 9 and 10) will go on to argue that, in the case of London, far from adopting the language of 'not blaming the victim', we should not be exonerating the local.

However, within this dynamic of differential relations London's own growth, and the wider spread of growth around it, constantly threatens to undermine itself. London can no longer function without 'special measures'. As well as London weighting, it requires a plethora of programmes to attract and retain 'key workers', interventionist programmes to provide 'affordable housing', measures to support public services periodically in crisis because their workers can no longer afford to live in the area they serve. . . . More strategically, vast new housing estates are planned for the South and East to cope with predicted growth. But special measures are really only a stop-gap approach: patch and mend. Some of the measures, caught up immediately in the dynamic of wider market forces, become inadequate almost as soon as they are implemented. Unless the current inter-regional dynamic is changed these demands will continue, and will continue to escalate. Indeed, to accede to them is only in the medium term both further to fuel the flames and to exacerbate the North–South divide. Is this a responsible capital city?

This patch and mend approach results from a wider politics of responding with far more alacrity to the problems of growth, such as congestion, than to the needs of regeneration, for instance in the North. If the stated policy aim of a more regionally egalitarian country is genuine, then it demands a different strategy.

First, a strategy for greater regional equality must be concerned not only with some post-hoc redistribution but with countering the production of inequality in the first place. One obvious measure here is the institution of a regional audit on all policies and especially those that might seem to have no regional dimension at all. It must, however, be an audit that is public, and that is responded to.

Second, there must be measures to address the concentration into one corner of the country of social, economic, cultural and political power. Regional inequality does not begin and end with such measures as per capita income, GDP, or unemployment, important as these are. It both reflects and is fuelled by a massive democratic deficit beyond the capital and the self-referential elites that congregate there. Nor is this democratic deficit simply solved by devolution of regional or local affairs. The real issue is the decentering of the centre (Amin, Massey and Thrift, 2003).

Third, there should be an explicit overall guiding view of the geographical structure of the country. And this means nationally led initiatives. One obvious sphere for such initiatives is transport. But it is also more than that. Instead of goading the regions to compete against each other, and in the process most likely come up with very similar strategies (*every* region aiming to be 'the national centre for the knowledge economy', or whatever), there need to be negotiated national strategies of concentration. Not only would this be more effective in terms of public expenditure, but only through this kind of concentration might it be possible to achieve constellations that could be counter-attractions to London/the South-East.

Fourth, a similar approach is needed at supranational level. The EU can wield sufficient power to face down the multinationals and to prevent, for instance, inter-country competition for investment.

Fifth, it is simply inadequate to ignore the impact of London's growth on other regions, or to insist that this impact is simply benign. London's growth has impacts in a variety of ways: through simple *size*, and the consequent expenditure demands (e.g., the costs of congestion that that generates); through the *concentration* of certain things into London, and the consequent relative deprivation of regions elsewhere; and through the *nature* of the growth, for instance its current overwhelming focus on professional jobs.

But there is one issue above all. Inequality between regions is bound up with national levels of inequality and with the dominance of the Anglo-Saxon model of economic growth. The increases in regional inequality are intimately related to the breakdown of the social democratic settlement. Two elements in particular stand out. First, the increased dominance of the City constellation and all that has gone along with that. This has both directly exacerbated the North–South divide and, in its intimate relation to deregulation and marketisation, has had further, also regionally unequal, ramifying effects. Second, the new elite of the very rich. As Krugman (2002) argues (see chapter 2), it is the abandonment of previous social norms rather than any simply economic or 'market' justification that accounts for the skyrocketing of top salaries. Either limits must be set (Kettle, 2006) (and here the supranational level is again important) or a more progressive tax structure must be introduced. The social norms of the social democratic settlement 'were reflected in marginal tax rates. With high marginal tax rates it made little sense to pay high salaries . . . . Once top marginal tax rates were reduced, as in Anglo-American economies, salary increases were largely translated into post-tax income increases for their recipients' (Dunford, 2005, p. 159). The generalising, and social acceptance, of this lies behind not just the national but also the inter-regional increases in inequality. The two interact. Since the 1980s increasing levels of national inequality have been one important generator – through the concentration of the richest echelons in London – of regional inequality. But that spatial organisation of differential wealth has its own effects in each place – in some regions through the concentration of poverty, in London through the impact of inequality within the region. And, in reverse as it were, through such mechanisms as the differentiated geography of house-price rises, regional inequality feeds back into national levels of inequality. The geography of a country is part and parcel of its very character and functioning, and it is in that integral context that it should be addressed.

Part III

THE WORLD CITY IN THE WORLD

8

GROUNDING THE GLOBAL

Local places all around the world these days claim to be global. It is the aim of city governments on every continent. It is, however, a claim that, if taken seriously, hints at a problem with some of the dominant imaginaries of local and global. It also, potentially, points towards a very different politics.

So often, and in spite of the fact of ritual denunciations of such practices, 'local' and 'global' in discourses both intellectual and political are counterposed. In one version of this counterposition, the local is the seat of authenticity – real, grounded, the sphere of everyday life – with the global functioning in contrast as an abstract dimension of space. In other versions the local is the produced outcome, the global the sphere of the forces that produce. So, on this reading, the local is a product of the global and, in counterposition, the global is figured as always emanating from elsewhere. We saw, in Part I, some London examples of this – the global imagined as always arriving from beyond even at the same time (in the same paragraph) as global citydom was being proclaimed. This is a manoeuvre that obscures the real geographies at issue: 'neoliberal globalization conjures up the image of an undifferentiated process without clearly demarcated geopolitical agents or target populations; it conceals the highly concentrated sources of power from which it emanates and fragments the majorities which it impacts' (Coronil, 2000, p. 369).[1]

This is a geographical imaginary which can be, and is, mobilised by both the political right and the political left. At its crudest it can function as support for the rejection of any arrivals from 'outside', be that in-migrants from the global South, perhaps, in the case of the political right, or multinational corporations in the case of the political left. At its most general it can buttress a political cosmology in which the very terminology of 'localness' carries with it an implication of goodness and warmth. Further, the understanding of the local place as *product* (at the receiving end) of global forces can slide very easily into an imaginary of the local as *victim* of the global. And this in turn can lend itself to a politics in which the aim is to *defend* the local against the global. This is a geographical imaginary with a host of roots and resonances. It draws upon a widely understood distinction between space as a modern, scientific and universal dimension and place as the locus of tradition and specificity. It is heavily interwoven with a differentiation between agencies, and even genders. As Escobar writes, 'the global is associated with space, capital, history and agency while the local, conversely, is linked to place, labor and tradition – as well as with women, minorities, the poor and, one might add, local cultures' (2001, pp. 155–6). It works, too, from a background assumption of space as always already territorialised.

There is a host of ways in which such an imaginary is open to criticism. First of all there is the in-principle argument that it tends to harbour a spatial fetishism in which particular geographical forms or scales are understood to carry a given political content – for some: local good, global bad; for others: the other way around. It is not spatial form in itself, but the particularities of the social construction of that form in any specific instance, that should be the focus of political evaluation. This will be evident in what follows. Second, in empirical practice, the new geographies of global-isation give the lie to all that (even if older imperial geogra-phies could not). This is pre-eminently the case in a place like London. In such a place it is not only that the local is not simply a product of the global, but that the global itself is produced in local places. This is an argument that has been made forcefully by Sassen (1991, 1998, 2000). The 'global' forces that have their effects in London by no means always have their origins elsewhere. Manifestly, this local place is not purely a 'victim' of the global. Here, too, it is clear that the frequent equation between the local and the everyday simply cannot hold, and this is the case in a thousand – though varying – ways of most groups within this place. It is evident, too, in the trading and the transactions, in the flows of products and of cultures, that the global is as material, and practised, and grounded, as the local is usually singled out to be. In a place like London it is plain that a serious politics cannot restrict itself to a defence of the local against the global.

Such a politics needs to address the global positioning of places. If space is conceptualised relationally, as the product of practices and flows, engagements, connections and dis-connections, as the constantly being produced outcome of mobile social relations, then local places are specific nodes, articulations, within this wider power-geometry. It is this relational constitution that renders so patently inadequate that rhetoric of regions and countries as autonomous enti-ties, able to be held up for approbation or disapproval for

'their' success or failure. Moreover, different places are
formed of *distinct* nodes of relations, distinct positionings,
within the wider global spaces. Each place is a different articulation of relations and connections, in some of which it will
be in a position of relative control, influence and power, and
in others of which it will be comparatively powerless and
subordinated. The degrees of 'victimhood' to forces emanating from elsewhere will, in consequence, vary. In some places
there may well be some purchase, at the local level, on so-
called global forces – some possibility for active intervention.

That spatial imaginary in which the local must always be
defended from the incursions of the global has, on the left at
least, developed largely in situations which are on the receiving end of forces that seem to arrive unwanted from elsewhere and to bring havoc in their wake. Such places may be
in the global South (see Escobar, 2001) or in shattered manufacturing communities in the countries of nineteenth-
century industrialisation. Yet even in such places important
work has been under way to counter that 'victimhood' relation between local and global. Thus, as Gibson-Graham
writes, 'Globalization discourse situates the local (and thus
all of us) in a place of subordination, as "the other within" of
the global order. At worst, it makes victims of localities and
robs them of economic agency and self-determination'
(2003, p. 50), and she urges us to 'imagine what it would
mean, and how unsettling it would be to all that is now in
place, if the locality were to become the active subject of its
economic experience' (ibid.). In an attempt to actualise this
imagination she develops what she calls 'an ethics of the
local' (the title of her article). Here it is the inevitable impos-
sibility and incompletion of any single global order ('the
local cannot be fully interior to the global'; ibid.) that provides the necessary room for manoeuvre: 'Such an ethics is
grounded in the necessary failure of a global order, which is

the negative condition of an affirmation of locality' (ibid.). The approach adopted builds on Foucault's notion of self-formation as an 'ethical subject' (Foucault, 1985, p. 28) and of 'modes of subjectivation' through which such a subject is supported. This is a research project thoroughly embedded in political engagement, in the Latrobe Valley in south-eastern Australia and in the Pioneer Valley in Massachussetts in the USA.[2] 'In both these regions, globalization sets the economic agenda – we are all being asked to become better subjects of capitalist development (though the path to such a becoming does not readily present itself) and to subsume ourselves more thoroughly to the global economy' (Gibson-Graham, 2003, p. 56). As Gibson-Graham summarises:

> In the discourse of globalization, the economy is something that does things to us and dictates our contours of possibility. It is not the product of our performance and creativity. Globalization discourse represents localities as economically dependent, not so much actors as acted upon, receiving the effects of economic forces as though they were inevitable. In the face of this representation, the urgent ethical and political project involves radically repositioning the local subject with respect to the economy. (Ibid., p. 54)

In this case, the radical repositioning concerns the assertion of the significance of other social relations of production as against the presumed total dominance of capitalism. Moreover, these local places are globally positioned in a nexus of relations that sets them as more (if not, as Gibson-Graham so cogently argues, entirely) on the receiving end than as leading protagonists in the global capitalist economy. But what of other places? The contrast with the global positioning of London, and of all those other places claiming and/or aiming to be 'world cities' too, is clear.

The irony is that in the United Kingdom the dominant discourses work precisely in the other direction. A good number of UK cities – Manchester, Birmingham, Newcastle, Glasgow, Liverpool – and regions have in their time been local places dominant within global imperial relations. Currently, however, they are set within, and are internally constituted largely through, power-geometries that position them in comparatively subordinate ways within the wider economy. Compared with London and the South-East they have more claim on that refrain of 'local place as victim of the global'; they have relatively little relational power. Some parts of those regions and nations more resemble the Latrobe and Pioneer valleys of which Gibson-Graham writes. And yet, as was seen in Part II, it is these regions that are being urged to stand on their own feet and compete. With so little room for manoeuvre and potential leverage, it is they that are being instructed to bootstrap themselves into renewed growth. In contrast, London and the South-East are just accepted as having to grow, as being unable to resist the global forces that heap upon them ever increasing wealth (for some) and economic activity (of a particular kind).

London, evidently, is precisely one of those places that are in a position of relative power within the global economy. It is one of those places in which the current form of the global has been imagined and through which it is constituted. The relational constitution is crucial. My concern here is to stress the political importance of recognising this. In places like London it is more important even than elsewhere to move beyond an imaginary in which the local is victim of the global. It is also important for reasons that are different from those in other places. For the assertion of local agency in a global context, here, arises not only from the need to reinvigorate and reinvent the economy *within*, but also from the need to recognise this place's implication in the production of the global itself, and what that means for other places (such, indeed, as the Latrobe and Pioneer valleys). The fundamental point here is that there is a need to escape from that spatial imaginary which forever leads, within the context of the production of neoliberal globalisation, to the exoneration of the local place. Rather what is needed is a politics that is prepared not just to defend but also to *challenge* the nature of the local place, its role within the wider power-geometries. What is needed is a politics that recognises, rather than persistently deflects, the role of the local in the production and the maintenance of the global.

In the days after the bombings in July 2005, the defiant celebration of London's identity focused on that aspect of London-global-city that is its housing of so much in the way of difference. As Gilroy has urged us more fully to recognise, there is in the cities of the UK, and especially in London, a lively and assertive 'convivial culture', a demotic cosmopolitanism, which thrives in the face of that other national narrative, of 'post-imperial melancholia'. It is a street-level reinvention of identity. It was these 'chaotic pleasures of the convivial postcolonial urban world' (2004, p. 167) that were at the centre of the self-conception of London in the late summer of 2005 and which run as a strong current of identification in the city more generally. Gilroy, with a fine appreciation of the political significance of recognising and working with multiple trajectories and specificity, and with a refusal to convene such coexisting specificities into temporal sequence, writes in the final paragraph of his book:

> The recent history of Britain shows that it does not lag behind the United States in racial politics but has embarked on an altogether different path toward the goal of multicultural democracy. . . . I hope it does not sound melodramatic

to say that the future of Europe depends upon what can now be made of that legacy. (Ibid., pp. 167–8)

Robins, not dissimilarly, and evoking London as 'that great provocation to the clarity and coherence of British national culture' (2001, p. 77), also argues that this aspect of its internal identity can/should have wider implications:

> the point about London is precisely that it is *not* a nation – but a city, a metropolis. And, as such, it allows us to reflect on the cultural consequences of globalization from an other than national perspective . . . to . . . open up some alternative cultural and political possibilities . . . I would argue that now, in the context of the new order of cultural complexity being brought about by the processes of globalization, London provides a crucial intellectual framework for British people to re-think and re-describe their relation to culture and identity. (Ibid., pp. 86–7)

All of this resonates too with Ken Livingstone's and others' reflections on London after the bombing: 'This city typifies . . . a future where we grow together and we share and we learn from each other' (GLA, 2005b). It is an argument that, while crucial to any analysis of London-world-city, is by no means unique to it. Writing of cities in general, Bender asks: 'Might the city – in its metropolitan form, acknowledging its embeddedness in structures larger than itself – again be the place and means for thinking oneself into politics and acting politically in the circumstances of our time?'(1999, pp. 37–8). What this points to more generally is an aspect of the wider political potential of the negotiation of place.

All of these understandings figure local place as open to the wider world, as articulations of a multitude of trajectories. They are understandings of place as generous and hospitable.

They are definitively not about any closure of the local to the global, nor about a figuring of 'local' as cosy and good against a global as out there and threatening. Can such an understanding of a wider global positioning be extended to that other incarnation of London as global city – as inventor and protagonist of deregulation and privatisation for instance? And can the relations which connect to that wider world be traced back to their 'other ends' elsewhere?

In that 'elsewhere' both inequality and absolute poverty are increasing. In 2005, the annual *Human development report* produced by the United Nations documented, alongside some small achievements, 'an unprecedented reversal': eighteen countries, with a combined population of 460 million, registered lower scores on the human development index than in 1990 (Elliott, 2005). 'This year', said the report, 'marks a crossroads.' Replicating the expansion of the small group of very rich in London, at the world level the richest 500 people own more wealth than the poorest 416 million. Even some of the poorest countries are seeing the emergence of a stratum of super-rich (Kundnani, 2006). Such inequality, and absolute poverty, are intimately connected to the tensions within London that were analysed in Part I, and to the inequalities there as well. This is not a claim for simple, singular causes – the forces behind the situation described by the UN are multiple and complex. However, a recognition of that complexity should not be used to deny the connections altogether. 'London', through the terms of its recent reinvention, is implicated in this.

London 'has reinvented itself' with and as an integral part of the neoliberal model in its stark Anglo-Saxon variant. It is not the case that this 'neoliberalism' is some undifferentiated force that has swept all before it, nor that all city governments everywhere have interiorised neoliberalism into city policy and used their urban bases as laboratories for neoliberal experiments, as some would have it. But London's role in this global model is undeniable. Its acceptance of the pressure to compete with other places for global-city status is in itself a reinforcement of neoliberal ways (Peck and Tickell, 2002) and has as one of its outcomes the need to attract and provide for an already more than comfortable elite. As the Introduction noted, the question should at least be asked as to how much of London's current 'success' is a product of the selfsame forces that result in such poverty elsewhere.

This is not a question asked, for instance, by the current *London plan* (GLA, 2004b). Having established a general context in which London is in the grip of wider forces over which it appears to have no say, the global dominance of the City as a financial centre is presented as a simple achievement. There is little questioning of what is actually *done* here, or of what this city as a financial centre actually represents, in terms of a shift in economic doctrines and the establishment of a new class hegemony. There is no following of relations around the world to ask what they are responsible for. There is no questioning of any possible connection between this financial power and inequalities and poverty elsewhere. Indeed, on this issue its analysis of 'relations with elsewhere' is pervaded rather by anxiety about *competition* with other places.

There is a host of policies emanating from this elite, and upon which London's growth now seems so firmly predicated, that would merit interrogation in this way. Among the causes of the consequences, in poverty and inequality around the world, for which neoliberalism has been most held responsible, the deregulation and 'the financialisation of everything' stand out as being the most critically debated (Harvey, 2005; Held, 2005). It is precisely these which have been central to London's revival. Yet there is no wide public political debate about the global implications of this aspect of its world citydom. There is also a host of more subtle and particular ramifications. For instance, as was seen in Part I, the effect through the housing market of the presence of the very rich, and indeed of the growth of professional strata more generally, has been to contribute to the difficulty of London's reproducing itself. Public-sector workers and lower-paid private-sector workers are hard to recruit. One result of this is that London is seriously dependent for its normal functioning on labour from elsewhere (see chapter 2). Some of these workers come from other parts of the country – some of the implications of this have been explored in Part II. But many others come from Eastern Europe and the global South. London, just to keep itself going, is dependent, for instance, on nurses from Asia and Africa. These countries can ill-afford to lose such workers, and they have paid for their training. So India, Sri Lanka, Ghana, South Africa are subsidising the reproduction of London. It is a perverse subsidy, flowing from poor to rich. It is, moreover, a flow that is both fuelled and more difficult to address as a result – precisely – of the increasing commercialisation/privatisation of health services at both ends (Mensah, Mackintosh and Henry, 2005). This raises fraught and complex political questions (see chapter 10), yet it is not even at present a live political debate among Londoners. Even as Londoners, rightly, celebrate the arrival of such workers as part of the great ethnic mix, they (we) do not pause to follow those lines of connection out around the rest of the world to enquire about the effects they are having elsewhere.

Gilroy applauds the vitality of the emerging convivial culture as a challenge to the heretofore nationally pervasive mood of 'post-imperial melancholia', and this is absolutely right. But the financial City and the constellation of interests and social forces that surrounds it are by no means melancholic. Those who are at the heart of (this aspect of) London's claim to global citydom are triumphant and celebratory, as they pick up and build upon the threads of an older imperial order. By trading on long-established inherited links and connections, and through that discomforting complicity whereby even the old materialities, of wood-panelling and dress, can be reworked as heritage-based reassurance, a new imperial order has taken hold. And London (a part of London) is once again at the centre of it.

London (and the same could be said of many cities) is no place in which 'the local' can be simply defended against the global. Rather, in consideration of the facts of the global, it is more appropriate to *challenge* the nature of (some aspects of) this local place. Likewise, in all those places where contests are at present under way against proposals to reinvent them as global cities, the challenge could be made not only on the grounds of what that will do to the local place itself (the policy focus on new elites, the promise of trickle-down, the destruction of older 'non-conforming' areas), but also on the grounds of questioning the nature of the role/identity/effects of such global citydom, were it to be achieved, on the rest of the planet. This is a reimagination that demands thinking beyond the normal territorialisation of electoral politics, so inadequate in a world of flows. It requires a more outward-looking politics that seeks to address that wider geography of place and to ponder what might be thought of as the global responsibilities of (some) local places.

9

IDENTITY, PLACE, RESPONSIBILITY

It is nowadays increasingly accepted, certainly on the left broadly defined, that there is a need to be severely critical of the old British imperial order. Indeed the dawning realisation of some of the horrors that it had entailed has been, Gilroy argues, one of the elements contributing to the post-imperial melancholia. There have been some 'apologies' for past actions, there are ongoing arguments over restitution, and there is a variety of concrete (and often contested) attempts at recognition of complicity, acceptance of implication. At a civic, urban, level the Liverpool Museum addressing that city's past involvement in the slave trade is just one example.

One of the notions that arises here is that of extended responsibility – that is, a responsibility that is not restricted to the immediate or the local. In the case of the old imperial order, the crucial dimension of extension is temporal; the question concerns the nature of present responsibility for actions committed (by others) in the past. One way of addressing this question has been suggested by Gatens and

Lloyd in their thought-provoking book *Collective imaginings* (1999). Their concern is to think about the nature of collective responsibility, in present-day Australia, for white Australians' past actions towards Aboriginal society. The notion of responsibility they proffer has extension (in the sense outlined above) and is relational, in the sense that it derives from constitutive relations with others. They write, in relation to their own concern: 'In understanding how our past continues in our present we understand also the demands of responsibility for the past we carry with us, the past in which our identities are formed. We are responsible for the past not because of what we as individuals have done, but because of what we are' (1999, p. 81). This understanding of responsibility, in other words, poses it as deriving from those relations through which identity is constituted. A first question, then, is: can this extension of responsibility over the temporal dimension be paralleled in the spatial and in the present? For just as 'our past continues in our present' so also is the spatially distant implicated in our 'here'. Is it possible, then, to draw on this notion of responsibility in the context of a local place such as London in the *new* imperial order *now*?

Gatens and Lloyd's conception of responsibility depends upon an understanding of the relational construction of identity. The reworking of concepts of identity, away from the billiard-ball self-constitution of the isolated individual towards an open, processual and mutually constitutive understanding that has characterised social and cultural studies in recent years, has been paralleled in geography by a reconceptualisation of the identity of place. Such a reconceptualisation necessarily entails that the spatiality, as well as the temporality, of identities and subjectivities is something of consequence. Identities are, constitutively, elements within a wider, configurational, distributed, geography. And that raises a second question based on Gatens and Lloyd's proposition: the question of the real geography of relations through which any particular identity is established and maintained. For it is from those relations that would spring a geography of responsibility.

Insofar as this spatiality of identity has received attention, the focus has been overwhelmingly on the internal structuring of identity. There has been much consideration of the internal multiplicities, the decenterings, the fragmentations of identity and so forth.[1] And such arguments have been important theoretically and politically in grappling with issues, for instance, of essentialism. Likewise, in relation to the identity of place, the emphasis has been on exposing and exploring the hybridities within, the global within the local, the issue of hospitality, the strangers within the gate (London's internal mixity). And that, too, is important, both intellectually and politically.

There is, however, another side to the geography of the relational construction of identity, of a global sense of place. For there are also the relations that run *outwards*, the wider geographies through which identities are constituted. The strangers that remain with*out* the gates. To consider these would be to translate Gatens and Lloyd's concept of responsibility from extension in the temporal dimension to extension in the spatial. It raises the necessity for a wider, distantiated, politics of place.[2]

However, acknowledging responsibility for present wrongs, including those distant in space rather than in time, poses rather different challenges. Most evidently, it involves not just compensation for positions that are already unequal but at least some degree of address to the *production* of those positions. And this is in some senses a tougher claim. Yet temporal and spatial should not be counterposed. It is central to Gilroy's argument, for instance, that it is necessary more fully to recognise the iniquities of history, and to use that recognition in order to go beyond them in the present (see also Hall, 2000). And he too, as in the case of arguments here specifically about London, is concerned with that shift from the old imperialism to the new:

> Instead of reinflating imperial myths and instrumentalizing imperial history, I contend that frank exposure to the grim and brutal details of my country's colonial past should be made useful: first, in shaping the character of its emerging multicultural relations [that is, in the terms of the argument here, its internal identity], and second, beyond its borders, by being set to work as an explicit challenge to the revised conceptions of sovereignty that have been invented to accommodate the dreams of the new imperial order [that is, in the terms of the argument here, the 'external' relations constitutive of identity]. The revisionist ways of approaching nationality, power, law, and the history of imperial domination are, of course, fully compatible with the novel geopolitical rules elaborated after 9/11. They have also been designed to conform to the economic machinery of weightless capitalism [so-called] and work best when the substance of colonial history and the wounds of imperial domination have been mystified or, better still, forgotten. (Gilroy, 2004, p. 3)

It is precisely on a mixture of mystification and forgetting that the central plank of London's reinvention, and this aspect of its current claim to world citydom, has been built. Perhaps a greater recognition of that past could now form one way into an acknowledgement of our extended and relational responsibilities for the spatially distant present. Gilroy's call is for, precisely, a greater commitment to 'translocalism', a 'translocal solidarity'.

Yet London's identity, its ability to be and continually to become what it is, is built on far more than the 'finance and business services' and the neoliberal constellation at the centre of the new imperium. Indeed, even in simply economic terms its identity is above all 'diverse' (Part I). And the relations upon which it feeds go far beyond the economic, to take in all aspects of the cultural, social and political. In obvious material terms London's existence depends on daily supplies from around the planet and, at the other end of the process as it were – its production of waste, its emission of carbon – its footprint is also geographically extensive. 'Ordinary Londoners', as well as the significantly wealthy, share in the responsibilities imposed by this identity. The 2005 UN *Human development report*, as well as highlighting the extremes ('the richest 500 people', and so forth), also points out that Europeans spend more each year on perfume than the $7 billion needed to provide 2.6 billion people with access to clean water.

Young has addressed this more ordinary implication in an article subtitled 'sweatshops and political responsibility' (2003; see also Young, 2004). Again the concern here is with responsibility at a distance, and in this case the dimension of extension is explicitly spatial – the immediate relevance for her is that geography which holds together consumers in the United States of America and workers in sweatshops in the global South. Young's concern is to move, as she puts it, 'from guilt to solidarity' (her main title). In the case of guilt, she argues, if some are culpable others are thereby absolved. However, in the case of political responsibility, this is not so; there is no isolatable perpetrator. Rather there is a chain of ordinary actions: 'many harms, wrongs, and injustices have no isolatable perpetrator: they result from the participation of millions of people and institutions' (2003, p. 41). Young does not tie responsibility to identity as do Gatens and

Lloyd. She specifies it, rather, in terms of participation in structural processes – those structural processes that lead from daily lives (in London or any other Western city, for example) to global inequality.

However, in this distinction between guilt and political responsibility, Young engages with the different implications of extension in space on the one hand and extension in time on the other. Guilt, she argues, is usually taken to refer to an action or event that has reached its end. For that reason it tends to be backward looking; it is about the past. 'Political responsibility looks forward rather than backward. Blame and praise are primarily backward looking judgments. They refer to an action or event assumed to have reached its end. The purpose of assigning responsibility as fault or liability [the guilt model] is usually to sanction, punish, or exact compensation. . . . Political responsibility doesn't reckon debts, but aims at results' (2003, p. 41). On this understanding, accepting responsibility for the old Empire would be very different, and separate, from acknowledging complicity in the new. However, past and present are intimately connected and in, at least, two ways. First, Gatens and Lloyd's approach to responsibility through identity avoids any possibility that the past can be dispensed with through sanction, punishment or compensation. On their formulation, the past (in their case the past treatment of Aborigines) cannot be rendered into a closed book, and this is so not because the maltreatment continues into the present. Rather, the issue is not closed and they/we are still implicated (responsible) because those past actions, by others, are part of what makes us what we are. In just the same way, the 'old Empire' of the past has provided a foundation for the reinvention of London as new global city. Second, as Gilroy argues, any present attempt to build on that convivial culture, and support the incipient demotic cosmopolitanism, which in London as perhaps nowhere else exists in intimate relation with one hearth of the new empire, will be greatly strengthened by a 'frank exposure to the grim and brutal details of my country's colonial past' (2004, p. 3). Can that vibrant, ordinary, cosmopolitanism that now opposes the post-imperial melancholia, also oppose – through acknowledgement of its own position within it – the new imperialism?

There is, moreover, one final step in this argument about responsibility, and one which draws the street-level convivial culture itself into the structure of implication. Young argues that there is still a further distinction between responsibility over temporal distance and responsibility in the spatially distantiated present.[3] This is that reparations for past events single out those events as having been 'abnormal'. 'In a blame or liability conception of responsibility, what counts as a wrong is generally conceived as a deviation from a baseline. Implicitly, we assume a normal background situation that is morally acceptable, if not ideal' (2003, p. 41). 'Political responsibility', on the other hand, 'questions "normal" conditions' (ibid.). 'A concept of political responsibility in relation to structural injustices, on the other hand, doesn't focus on harms that deviate from the normal and acceptable, but rather brings into question the "normal" background conditions' (ibid.). The identity of London is constructed and maintained through the same relations (the 'normal' workings of neoliberal globalisation) that produce the conditions described by the UN *Human development report*. It is 'normality' here that should also be put under question.

'Local place', then, can be one potential basis for political organisation around responsibilities of this sort. A critique of local place as simply defensible space, as exclusivist, bounded, romanticised, does not imply that place is not a potential basis for political organising. Indeed, as we shall see, place-based politics as suggested here in itself both reinforces, and plays on, the notion of place as *un*bounded and potentially sparks new lines of (productive) internal debate. What is being suggested here is a networked, practised, internationalism. It does not necessarily stem from abstract and/or universalistic claims (for instance about 'humanity in general'). Yet it definitively goes beyond the local; and it challenges, by reworking, the notion of 'particularism' (Featherstone, 2005). It is a local internationalism that challenges the dominant geographical imaginary which understands the world in terms of scales and nested hierarchies. This relates, again, to bigger issues about politics and agency. The view of the world that understands its politics solely in terms of big binary divisions (solely us versus them) is more than likely, also, to harbour an implicit geographical imaginary of global versus local, and to associate 'us' and 'local'. It is a manoeuvre that both positions us/the local outside the dominant structures of power (the exoneration of the local), occluding the inevitable implication in these structures that is addressed by Gatens, Lloyd and Young, and closes down the space of political agency. A more complex geography of politics opens up both. The local place becomes one (though only one) potential arena for action to change the global. This is, moreover, a complete reformulation of the usual scalar notion of 'local politics' (often in Europe encapsulated in the term 'subsidiarity') in which the nation deals with big, national and international, issues, while local areas are told (patronisingly, for these issues are presumed not to be important – though they are) that they can get on with the positioning of bus stops.[4] Local internationalism ignores such hierarchical presumptions. It cuts right across the scalar geographical imagination that supports the discourse of subsidiarity. Local authorities should have their own 'foreign politics', in the sense of enquiring into and taking responsibility for the wider implications of their places. And this is a matter not only for local states but for local places in a wider and more grass-roots sense. This could contribute to a more grounded (and alternative) globalisation that based itself firmly in the material juxtapositions of place while at the same time insisting on an acknowledgement of openness. Moreover *within* place the same point applies: the issue is not only (though it is most importantly) one of challenging the big battalions – in London, the financial City for instance. 'Ordinary Londoners' are implicated too.

Moreover, rather than taking the identity of place for granted, as a given, as do so many campaigns to defend the local, this kind of local politics is more about throwing up the challenge: what *is* this place? In that sense, as in Gibson-Graham's rather different settings, it is about an assertion of local relational agency, and the question of an ethics of place, in an actively inventive manner. Moreover the issue of place identity can both bring an immediately available focus and tie in what can seem very general claims to particular practised relations. It can be rooted in the realities of a recognisable interdependence. It can provide a *locus* for campaigns and one that, on the arguments above, can move away from the ground of individualised culpability towards the terrain of collective responsibility. And if cities are crucial to neoliberalism, as is so often argued, then battles precisely over this role must be potentially significant to any challenge to its hegemony (see also Mitchell, 2004). The politics and economies of cities, and social struggles over them, are of crucial importance in defining the kind of world that is currently under construction.

This does, however, raise the question of the relation between the identity of place and the identity of individuals living within that place. That is to say: what are our responsibilities 'as Londoners'? – a question that can and should be asked about one's relationship to any place. Globalisation itself has made that question both more complicated and more urgent. Indeed the potential deracination that may be a product of mobility, especially where that mobility is differentiated socially, can both dramatically change the balance of forces in any negotiation of place and pose challenges to any notion of local democracy at all. The mobile gentrifiers are frequently accused of having no commitment to places which may figure as being no more than temporary bases in their globally peripatetic lives (and this in spite of the power which their very mobility – their ability to move on – can lend them). But the point is a general one. In this regard, I was startled, and interested, to be challenged at a recent conference. I had been presenting an argument like the one here, about the potential responsibilities of Londoners, and had identified myself as implicated in this responsibility. A thoughtful hand went up: 'But to me you're not a Londoner at all. You come from the North. *I've* lived in London all my life.'[5] Now, I harbour a real scepticism about some arguments for a strong relation between place and personal identity, especially where they depend upon longevity and rootedness rather than upon the more active notion of participation in the negotiation of place (see Massey, 2005). This in no way denies the feelings, and the political responses, of long-term residents caught up in the crossfire of globalisation (see chapter 2). However, as Livingstone said on that July day in 2005, people come here 'to call themselves Londoners', and indeed *most* 'Londoners' were either born elsewhere or retain some, often quite strong, threads of attachment to other places too. There is a weave of multiple allegiances that defies an either/or characterisation and which recalls that complex and intersecting notion of multiculturalism proposed by Saghal and Yuval-Davis (2006). Indeed this very dispersedness of attachment seems to be one of the characteristics contributing to the identity of the place itself. And, as we have seen (in the Introduction), pollsters find that a goodly part of this heterogeneous population do indeed identify as 'Londoners'. This returns us to that '*internal* geography' of relational identities, which is always multiple and hybrid. Not only is London's identity heterogeneous in this way but so also are the place-related aspects of identity of those who live within it – there is some part of our identities that is as Londoners. For after all, and following the argument set out at the beginning of this chapter, some part of what enables us to be what we are comes from the fact that we live here.

Furthermore, the fact that I am also a Northerner, and that so many Londoners are also of somewhere else as well, begins to change that question, alluded to in the Introduction, of 'where does London end?' Or maybe it is to pose the question in a different way. The imagination should not be of boundaries drawn ever wider, of ever bigger containers, or indeed of a new geography of scales. Rather what is at issue is a more dispersed geography, of relations and practices, and maybe even of identities: a different geography of identities within, as part of, globalisation.

Moreover what is at issue here is not only the way that the geographies of identity may be changed within globalisation, but also the *implications* of identity within that context. In identifying myself as a Londoner in an argument such as this, identity is serving not as the *assertion of a claim*, but as the *acknowledgement of a responsibility*. And that is very different. Perhaps such a reworking of identity/place/responsibility could feed into the extension beyond the borders of Gilroy's convivial culture into his translocal solidarity, his 'cosmopolitan solidarity from below and afar' (2004, p. 89). Maybe also it could add another aspect to that 'resubjectivation' that Gibson-Graham sees as essential to developing an ethics of place, extending it now also into an ethics of place beyond place.

10

A POLITICS OF PLACE
BEYOND PLACE

It is by no means impossible to envisage, in quite practical political terms, what such an outward-looking politics of place might look like. Indeed, in what follows there are examples of the implementation precisely of such a politics. They all defy the resignation that derives from that feeling of entrapment in bigger forces. They all take these forces on, each doing so in different ways, each of them presenting a challenge to the hegemonic imagination, whether that be of globalisation, or of local place, or indeed of simply what is possible. Each of them, in its own way, addresses that question: what does this place stand for?

In chapter 8 the question was raised of the dependence of London on, for instance, nurses from countries in the global South. This is a situation by no means specific to London, even within the United Kingdom. Indeed, the Medact project that was referred to in chapter 8 (Mensah, Mackintosh and Henry, 2005) was concerned with flows between *national* health systems (we shall return to this point). Such problems of the social reproduction of place are, however, particularly acute in global cities. Thus, Smith, writing of New York, recounts four events that 'succinctly captured some of the central contours of the new neoliberal urbanism' (2002, p. 427). One of these concerned the social reproduction of the labour force:

> in 1998, the New York City Department of Education announced that it faced a shortage of mathematics teachers and as a result was importing forty young teachers from Austria. Even more extraordinary, in a city with more than two million native Spanish speakers, a shortage of Spanish teachers was to be filled by importing teachers from Spain. Annual international recruitment of high school teachers is now routine. . . . Taken together, these events connote a deep crisis, not just in the city's education system but in the wider system of social reproduction. (Ibid., p. 428)

Such events, he argues, 'hint at much about the neoliberal urbanism that has been slouching towards birth since the 1980s' (ibid., p. 429). Once again, it is important to insist on national and local specificity. In part, this is indicated by the very geography to which Smith refers – New York is importing workers from Europe, from Austria and Spain. But also, the dependence of elements of urban reproduction on workers from elsewhere has characterised the UK since the arrival of the *Windrush*. It is by no means solely a 'neoliberal' phenomenon – indeed it has its roots in the older Empire. Nonetheless, and specifically in the case of London, it is a phenomenon exacerbated by the particular nature of the city's reinvention. Mackintosh et al. (2006) point to a recent sharp rise in health professional migration from low- and middle-income to higher-income countries – and particularly to the USA, the UK and increasingly Canada – and argue that this is not a temporary hump in overseas labour recruitment, but part of a sharp increase in international integration of markets for skilled labour. The market integration is being driven by changing technology, international competition for skilled labour, and in health by rising commercialisation of health care and vast global disparities in wages, working conditions, retirement prospects and the sheer scope for health professionals to do a good job. It is, moreover, particularly acute in London (RCN, 2003, cited in Mackintosh et al., 2006, p. 762). The element of this that is rising commercialisation of health care should be particularly noted, for it links the issue back, once again, to the terms of London's reinvention and its role in the current form of globalisation. And one consequence of the migration that this form of globalisation has stimulated is the worsening of the inequality between poor and rich countries (Mackintosh et al., 2006).

The brilliance of today's London, and the wider south-eastern region, is, then, dependent for its ordinary, daily, social reproduction on an array of workers from the rest of the world. This in-migration is indeed part of what contributes to the multicultural characteristics of the city from which this book began. It is another reason (apart from, or as part of, that freedom to be themselves) that people come to this city. The migrants are an element, also, in Gilroy's convivial culture. Many of them, such as nurses, bring high levels of skill which have been generated by investment by their countries of origin, maybe in the global South. It is, as noted, a perverse subsidy. Countries, and spokespeople, from the global South – including such powerful voices as Nelson Mandela – have made public their concerns and in some cases have pleaded for the flows to be constrained or even to be stopped.

This leads into extremely difficult political territory, and the difficulties are, precisely, spatial. At a general level there is the tension between on the one hand enabling individuals to realise their potential through movement and on the other hand the pressures of territorially based development strategies. For the political left, it can throw into apparent opposition two very different geographies of commitment. On the one hand, there is the commitment to an attitude of generosity and hospitality in relation to in-migration. On the other hand, there is the commitment at a global level to combatting inequalities between countries, and especially between 'the West' and the global South. Yet the first of these commitments works against the second. Unrestricted migration can result in increased inequality between countries. On the political right, and thereby posing another danger for the left, this is an issue that can all too easily be turned into one with an implicit, if not overt, racist inflection.[1]

One innovative way of cutting through these conflicting political spatialities is proposed by Mensah, Mackintosh and Henry (2005) (see also Mackintosh, 2007). This is completely to reimagine the relationship between flow and territory: to propose, in other words, another spatiality. Their specific concern is with the migration of health workers to the UK health system from Ghana, and their proposal is that the two health systems (Ghanaian and British) could be thought of as one system and that the United Kingdom could pay restitution to the Ghanaian element of that system for the perverse subsidy that currently flows in the opposite direction. Some of the issues that this raises will be returned to below. Before that, however, there are two other points to be noted.

First, the proposal for integrating the two national systems is not confined to the level of the state. This should, it is argued, be a grass-roots integration too, encouraging greater connections between trades unions and professional organisations in the two countries. This, then, were it to be implemented, would be a politics of engagement of 'ordinary Londoners' – in particular here trades unionists – as well as elected bodies. In this sense it would in some measure contribute to building an alternative globalisation to counter that which is currently hegemonic. It is moreover about the process of *construction*, not the prior assumption, of a grounded solidarity. Indeed Mackintosh reports both that the issue is only on the national agenda in the UK at all because of its being raised by UK activists, health trades unions and the Department of International Development, and that even the process of making the case for this politics has been generating a sense of social citizenship and solidarity around the interconnections of UK health services with African health services.

Second, however, as presently constructed this is a national-level issue. It is not the kind of political question normally addressed at urban level. There are two aspects of a political response to this. First, that neither the 1980s GLC nor the current mayorality have found this a major constraint in the past. London's voice has been raised over a range of issues that, while most certainly not 'normally' matters of urban policy, have in one way or another been of particular importance to the city. This is precisely part of the potential spatial reformulation of 'local politics' that was argued for in chapter 9. And this is the second point, that 'London's' voice is powerful. London can be an important player in national politics – indeed, as we have seen, through mobilisation of its 'global' status it already frequently is.

Let us return, however, to the proposal for the payment of restitution for the perverse inter-place subsidy currently flowing from Ghana to the UK, and to the wider proposal for an integration between the two health systems. From the point of view of the argument here the proposal is also interesting for the way in which it rearticulates geographies and

geographical imaginations. Firstly, the proposal takes the
globalisers at their word in their proclamations of openness.
It is, as Mackintosh et al. note, entirely in tune with a globalisation understood as the 'international integration of markets for capital, labour, services and goods' (2006, p. 757).
This is a straightforward challenge to the geographical imaginary of globalisation and thereby a way into the central
political argument. Secondly, it is a proposal that forces a
reimagination of place. It necessitates a recognition of interdependence, and of the inequalities within that interdependence.[2] It is a (potential) politics of place that looks from the
inside out. It recognises not just, as in the more usual formulation, the 'outside' that can be found within, but also – in a
certain sense – the 'inside' that lies beyond. It poses the
question of whether, in certain realms, we could imagine
(aspects of) other places as in a sense part of our own place,
and vice versa; or maybe live in the imagination of 'our own
place' as constituted through a distributed system – a kind of
multi-locational place. Thirdly, it poses in a different way
the potential for a politics *between* places, a politics precisely
of spatial (inter-place) relations that could be very different
from – and thereby a challenge to – that neoliberalisation of
inter-place relations highlighted by Peck and Tickell (1992).
It is the type of strategy that could be an element in a politics
that *linked* places in different positions within the wider
power-geometries of neoliberal globalisation – a (powerful)
form of inter-place solidarity.

Fourthly, this proposal is significant because it transforms
what otherwise might be conceptualised as *aid*, with all the
connotations of conditionality and charity and the power
relations thereby implied, into a matter of the fulfilment of
an *obligation* (Mackintosh, 2007). This again links into the
politics of geographical imaginations. Mackintosh makes the
important point that to generate a reverse flow as *aid* means

that it is imaginatively disembedded from economic relationships. Even while attempting to address inequality, it does
not admit of any implication in the *causes* of that inequality,
or to any political or ethical *responsibility* to reverse the existing perverse subsidy. Rather it is, astonishingly, perceived as
an act of generosity. The notion of restitution, in contrast,

precisely embeds the need for such a reverse flow in existing,
unequal, spatial relations. (In fact it is, as Diprose, 2002,
would argue, the countries of origin that have been being
generous.) It is a response, indeed, that accords with the
argument of previous chapters and that accepts responsibility
for the aspect of those relations that is the inequality from
which the richer place both benefits and, in part, takes its
identity. In other words, embedded within this proposal is
not only a challenging politics but also a radically different
geographical imagination. Moreover, adopting such a strategy would be to globalise in some way the local claim to
multiculturalism. It would be to begin to respond also to
some of the responsibilities that arise in consequence of that
aspect of the city's (global) identity.

The notion of this kind of outward-looking responsibility
is in fact present in some elements of the current plethora of
plans being produced in London. The *London plan* itself –
the overarching document (GLA, 2004b) – does on occasions recognise that London is a *cause* of what happens in the
wider world, that it does indeed have some agency in that
regard (it is not simply a local victim of the global) and that
it can and should take responsibility for its effects. This is
particularly the case in relation to climate change and environmental issues more generally: here it is stated that not
only must policy be directed to managing the impacts of climate change *on* London, but it must also work to reduce
London's own contribution to the *production* of that problem.
Thus, a 'key policy direction' is to 'Address issues of climate

change and ensure that the environmental impact of a growing London does not contribute to global warming' (2004b,
p. 10), and similar statements are made about waste. There is
a separate Waste Management Strategy, an Energy Strategy
and a Green Procurement Code, all of which embody similar
understandings.

The *Draft food strategy* (LDA, 2005) takes on directly the
issue of responsibility. Having celebrated some aspects of
food within the capital, it asserts: 'However, as many are
slowly becoming aware, there are problems associated with
this abundance; and London both contributes to and has
responsibility for some of these problems' (2005, p. i). The
overall 'Vision' has four components, with responsibility up
front, and is worth setting out in full:

In 2016, London's residents, employees and visitors,
together with public, private and voluntary sector organisations will:

- take *responsibility* for health, environmental, economic,
  social, cultural and security impacts resulting from the
  food choices that they make, and their role in ensuring that
  food and farming are an integrated part of modern life
- demonstrate *respect* for all the many elements involved in
  the provision of their food, and respect fairly the environment, the people, the welfare of animals, the businesses and others involved in providing their food
- be more conscious of the *resources* used in growing, processing, distributing, selling, preparing and disposing of
  their food, and be more engaged in minimising any negative impacts arising from this resource use
- benefit from the *results* of this effort, such that all Londoners have ready access to an adequate, safe, nutritious
  and affordable diet that meets their health, cultural, and

other needs, and better protects the environments in
which we live and those which we visit. (Ibid. p. iii)

The strategy addresses the issues of the context in which it
operates (such as the WTO, multinational corporations, the
EU), the power of market forces and the preferences of consumers. These things cannot be simply escaped.

Equally, however, the effect of a Strategy that simply accepts
such forces as 'given' is two fold. Firstly, it would not be able
to avoid responsibility. Secondly, it would represent a failure
to recognise market forces and consumer preferences as
dynamic and changing phenomena. Furthermore, policy
instruments, regulatory frameworks, information campaigns, targeted investment and political leadership can
actively shape and encourage the direction of change.

*London has the means to do this; and must accept the responsi-
bility to act.* . . . Equally, the responsibility for these changes
is widely distributed. (Ibid., pp.1–2, emphasis added)

This is impressive stuff. The development of international
supply-chains is mentioned, along with recognition both of
some of the inequalities embedded within them and of the
fact that London has in many respects benefited from these
developments. Eight stages in a supply-chain are defined
along with objectives for each. Thus, for example: 'Stage 2:
Processing and Manufacturing: London's role will be to
specify and expect high standards from processors based outside the capital that are supplying London, [and] to both
expect and support such standards within London itself'
(LDA, 2005, p. iii). Strong emphasis is placed on the range
of agencies that will have to support the resultant policies, on
the need for a lot of public campaigning within London, on
a range of initiatives to close the 'consistent "gap" between

attitudes *in theory* and behaviour *in practice*' (ibid., p. 64) and on a range of procurement policies including some mention of fair trade.

It should be stressed that the document referred to here is a draft and, at the time of writing, out for consultation. It is not yet known, therefore, what will result in practice. But this is an interesting document for the framework it establishes for policy – one which is both outward-looking and prepared to recognise local responsibility for the global.

The crunch point for such policies usually comes, however, when different political aims come into conflict. It is at such moments that the real priorities become apparent. There is a rather nice one in the *Food strategy* itself, where the intent to reduce food miles runs up against the fact of London's cultural diversity: 'Demand in London for ethnically- and culturally-specific food . . . is both much higher than elsewhere in Britain . . . and growing. Ensuring that all London's diverse communities continue to have access to culturally-appropriate food means that there are limits to the extent to which "local" food can meet London's needs' (LDA, 2005, p. 13). The two geographies of political intent come into conflict. Here, it seems, the demands of diversity win out, and this is made explicit. At other points of conflict, however, the matter is sometimes simply passed over. Thus, in the *London plan*'s consideration of transport strategy (intimately related to food miles) we find:

> The Mayor supports the development of a sustainable and balanced London area airport system, and recognises that further runway capacity in the South East will be required to meet London's needs. . . .
>
> . . . A sustained programme of development is needed if London and the UK are to compete effectively in the global and European economies. (GLA, 2004b, p. 110)

In other words, when it comes to the heart of the strategy for economic growth, and to competing with other places, the wider responsibilities are overridden.[3]

And yet there are things, sometimes small things, that are being done, or could be done, and not just by the local state alone. Most obviously and most crucially there could be a broader and more imaginative sectoral definition of London's claim to global-city status. The overwhelming prioritisation of finance and business services could be diluted. There could also be more explicit political recognition of the ways in which this current prioritisation actively poses problems for the rest of the city, including for other economic sectors (see chapter 2). It could, in particular, lay more stress on non-capitalist sectors, on 'restructuring for labour' and on the responsibilities as well as the demands of the capitalist sector (see GLC, 1985; Massey, 2001; Gibson-Graham, 2003). There could also be promotion of alternative forms of globalisation. The 1980s GLC gave encouragement in a variety of forms to the building of trades-union internationalism, aiding contact between workers in different parts of the world. London could join that growing alliance of city and regional groups that are refusing to go along with the provisions of the General Agreement on Trade in Services (GATS). There is a Fair Trade Town and City campaign that the city could join – such a move would build upon suggestions already contained in the *Food strategy* (see above) and might relate to public-sector procurement codes. But, again, it is not just that Londoners are major consumers. It is also that London is significant in the control of the production and trading of many of the commodities consumed. This is true, for instance, in the case of coffee. The GLC of the 1980s, recognising both this position and a point in the commodity chain at which it could effectively intervene, established Twin Trading, a wholesaling organisation, which continues strongly to this day and has been a significant actor in the wider fair-trade movement. There must be opportunities for targeted intervention within other commodity chains. These policies, around fair trade and GATS, are explicitly place-based and both in small ways challenge the nature of the trade and financial arrangements through which the current form of globalisation operates.

In fact, just as I was putting this book to bed, there came an announcement of a political initiative that would do just that. Following the visit of Hugo Chávez in the spring of 2006 there had already been joint cultural festivals between London and Caracas, but in September a proposal was put forward for a deeper relationship between the two cities. The proposal (which is still being negotiated) is for a barter arrangement in which Venezuela would send cheap oil to London in return for advice and experience in the areas of transport planning, housing, crime, waste-disposal, air quality and adult education. London would also agree to use the deal to promote Venezuela's image within the city. The oil, moreover, would be directed towards reducing the cost of public transport (buses) specifically for poorer people in London. The aim, in other words, is to be redistributive within both cities.

This is a truly imaginative strategy. It would bypass entirely the market relations of current globalisation and thereby present a challenge to them. ('Why could there not be more of this?' is the question it implicitly provokes.) It is an 'exemplary strategy' pointing to the possibility of alternative forms of globalisation. In that sense it is a small challenge, especially in its attempt at egalitarian fair-trading, to existing power-geometries. (Chávez, of course, is an explicit opponent of the Washington Consensus.) It changes, thereby, one small element of those global relations through which the identity of London (and Caracas) is constituted (and thus in a small way reworks London's identity too). It is, moreover, a direct challenge to the neoliberal mantra that cities (and places in general) must compete with each other; rather, they can cooperate. Like the proposal for restitution of perverse subsidies, through this inter-place solidarity, it links places in different positions within the wider power-geometries of today's globalisation.

There are also, and could be, grass-roots campaigns that target specific aspects of the world economy that are seen as being harmful and in which London plays a crucial role. The London Social Forum 'attempts to connect the local to the global in an alternative manner to the dominant neo-liberal model of which London as a major international centre of finance is a prime exemplar' (informal report to the Athens meeting of the European Social Forum, 2006). It would be good to see, for instance, a specific focus, in London-based campaigns, on making global links with struggles by fence-line communities in other parts of the world whose battles link back to companies in London. One obvious example could be the privatisation of utilities in the global South. Or, given London's huge financial role and participation in global offshore banking, there could be a strong representation in the city of the Tax Justice Movement. Oxfam GB (2000) argues that offshore financial centres are part of the global poverty problem: 'at a conservative estimate, tax havens have contributed to revenue losses for developing countries of at least US$50 billion a year' (ibid., Executive Summary). Wealth 'equivalent in value to one-third of global gross domestic product (GDP) is estimated to be held off-shore, and a large share of globally mobile capital makes use of tax havens' (ibid., p. 1). 'Tax havens and OFCs [Offshore Financial Centres] are now considered to be central to the operation of global financial markets. International banking activities, including the offshore currency markets (such as the Eurodollar market), are tightly inter-linked with the world of offshore finance' (ibid., p. 14) and 'many of the world's major havens are very much onshore. London and New York, for example, are both home to a substantial proportion of the world's offshore business' (ibid., p. 4). 'London, for example, has been the largest and most important centre of Eurocurrency operations since the 1950s. The favourable regulatory environment in London has ensured that international banks continue to carry out a large share of their international lending and deposit-gathering there, despite the rise of other financial centres. London is also the focal point of the Eurobond market' (ibid., p. 22, n. v). The aim of the Oxfam paper is to argue for a set of changes that will enable the 'hidden billions' syphoned off through this system to be released for poverty eradication around the world. As it argues: 'The UK is well placed to take a leading role' in this (and indeed it has made some moves). 'It also has a special responsibility on this issue as the UK is home to the City of London, a tax haven for some financial market instruments' (ibid., p. 17). More particularly, these activities have been at the heart of London's resurgence, and they constitute an integral part of its current identity. In one way or another (a point to which we shall return) the lives of Londoners are tied up in this.[4]

This, then, would be a campaign around particular aspects of the London economy, but with a focus on their global role, on the role that London plays in wider geographies. Another example is oil and gas. Together oil and gas account in one way or another for about a quarter of London's stock exchange; Shell and BP have major offices and headquarters in London; London is utterly dependent on oil. And a number of campaigns have focused on these facts and taken them as a starting point for wider arguments. London Rising Tide, a group campaigning around the root-causes of climate

Table 10.1   PLATFORM: celebrating 21 years of innovation, inspiration and impact

1. *Addenbrookes Blues* – solidarity work with striking hospital cleaners, through street theatre, backed by T.U.C. (1984–85)

2. Pioneering of safe spaces for difficult discussions, and the process of thinking as a vital part of activism: social sculpture (Free International University London, *Gog & Magog, Crude Operators*)

3. *Tree of Life. City of Life* – investigating London as an 'organism' over ten weeks in a mobile tent, with an exhibition at the Royal Festival Hall (1989)

4. *Still Waters* – imagining London as a city of rivers again, with its buried rivers restored to the surface (1992)

5. *Delta* – lighting a school with London's first micro-hydro turbine, using River Wandle water power (1993–96)

6. *Homeland* – deconstructing the process of international trade through the journey of a lightbulb across a continent, with London International Festival of Theatre (1993)

7. Transformative power of art and performance (*killing us softly*, *Carbon Generations*)

8. *Ignite* – publishing thousands of commuter newspapers challenging Londoners about the impacts of this oil city on the rest of the world (1996 & 1997)

9. *Carbon Generations* – investigating personal responsibility for oil dependency through a lecture/performance connecting family history and carbon consumption. (1997–98)

10. *Agitpod* – pedal-propelled, solar-powered, zero emissions video projection vehicle (1998 –)

11. *killing us softly* – day-long performance event involving research, poetry, music, video, discussion and a boat journey – investigating the hidden history of corporations and genocide (1999–2003)

12. *Internationalism & Solidarity work* – in Azerbaijan, Georgia. Turkey, against Pinochet; in former Yugoslavia; about Nigeria

13. *Gog & Magog* – detailed exploration of the complex web of organisations surrounding London's oil giants Shell and BP, using personal stories, guided walks and music (2000 –)

14. *Some Common Concerns* – a book presenting detailed analysis of the likely environmental and human rights impacts of the proposed Azerbaijan-Georgia-Turkey oil pipelines system (2002)

Table 10.1   PLATFORM: celebrating 21 years of innovation, inspiration and impact (*Continued*)

15. *Degrees of Capture* – a report exposing the capture of university research and teaching by oil corporations, and the detrimental impact of this on climate change (2003)

16. *SEA/RENUE* (1994 –) PLATFORM was the founder of RENUE, which, united with SEA in 2003, implements sustainable energy schemes in London and beyond.

17. *Unravelling the Carbon Web* (2000 –) working to prevent the environmental and human rights impacts of the oil industry – supporting those affected by oil in the former Soviet Union; exposing the moves to hand Iraq's oil reserves to transnational corporations; educating the public and decision-makers.

18. *Museum of the Corporation* (2004 –) a proposal for the world's first museum dedicated to the nature and impacts of the transnational corporation.

19. *The Body Politic* (2004 –) a pioneering interdisciplinary course for people committed to social and ecological justice, with Birkbeck, University of London.

20. *Remember Saro-Wiwa* (2004 –) initiating and co-ordinating a campaign for a Living Memorial in London to the inspirational writer and activist Ken Saro-Wiwa, executed for exposing the devastation of the Niger Delta by oil corporations.

21. *The Desk Killer* (2004 –) currently evolving from seven years of research, a groundbreaking book that investigates the history and psychology of corporate and bureaucratic killing.

Source: PLATFORM.

change, runs an annual 'Art not Oil' exhibition highlighting both the activities of oil companies and the sponsorship by those companies of art galleries located in London.[5] The radical London collective PLATFORM, combining campaigning and research, has among its numerous interventions often targeted this aspect of the city.[6] The project 'Unravelling the carbon web' (no. 17 in figure 10.1) explores and campaigns around 'the web of companies which constitute the international oil industry. The project focuses on

London, which has historically been the headquarters for two of the world's largest oil companies – BP and Shell – and is home to many companies servicing the oil industry' (Carbon Web Newsletter no, 2, p. 2).[7] Nigeria is central to the linked campaign 'Remember Saro-Wiwa: the living memorial'.[8] This was a public art initiative to mark the tenth anniversary of the execution of Ken Saro-Wiwa and eight colleagues for campaigning around oil extraction in Ogoniland in the Niger Delta. It was launched in spring 2005 in London's City Hall, with an opening speech from the mayor. But, precisely in the spirit of the argument in chapter 9, it is not a memorial that looks backwards – or rather insofar as it looks backwards it is in order to look forwards – to the continuing struggles in Ogoniland, for instance, and to the links between old and new empires. *The next gulf* (Rowell, Marriott and Stockman, 2005), produced as part of this campaign, emphasises the historical legacies and continuities (the 'gulf' here is the Gulf of Guinea):

> Four hundred years ago, the [Niger] Delta became a key element in the global economy, forming one of the three corners of the Atlantic Triangle. This triangle was built on the barter purchase of slaves in the Delta, their transportation to the plantations of the Americas, the production of sugar and tobacco on these plantations and the export of these tropical goods to the ports of Britain and Europe. London was pivotal in this triangle, profiting from the slave trade and coordinating the exports of guns and other items to the Delta as goods to facilitate barter. . . .
>
> The current trade in oil and gas, with the majority of Nigeria's output again crossing the Atlantic, appears to be recreating this triangle. Once again resources pour out of the Delta and guns flow in – though this time London shares its role with Washington. The last triangle was

broken by resistance in the villages of the Delta, rebellions on the slave plantations and the anti-slavery movement which began in London. Is it possible that the current triangle will be radically altered in a similar way? (Carbon Web Newsletter #2, p. 4)

The next gulf presents a map, 'The Niger Delta in London', which shows 'some of the companies and institutions related to Shell's operations in Nigeria' in 2005. If all the sources and links in the oil commodity chain and its multifarious supports were mapped, the centre of London would be crowded with references. An overlay of such maps, around a host of issues, would trace many of the elements and intricate interconnections in the City-centred constellation and the remoulded elite that was described in chapter 1. *The next gulf* has a similar map for Washington for the oil industry in Nigeria. The point is that similar maps of global implication could be drawn up for any place and for a multitude of sectors and activities. The potential of such maps, were they to be distributed widely, and made popularly visible, is great. On the one hand, they might dislocate a little the complacent imagination, make one stop, perhaps, for a moment and think what is being done here. And on the other hand, they are a quite different way of imagining that now perhaps all too easily invoked trope of 'the other within'. Here, in such maps, is being evoked the presence within this place of impacts on others beyond. And yet, without those impacts, this place would not be quite as it is.

There are many campaigns like the ones described here, and they are small, but part of their aim is to look beyond the local place, to trace its implications around the world. Some explore the ordinary invisibility of the way this happens, those chains of quotidian connection that Young implicates in political responsibility – the connections between panelled meeting-rooms, urbane gentlemanly discussions and decisions, and the havoc that may be wreaked elsewhere (see, for example, PLATFORM's 'killing us softly' and 'The Desk Killer', nos. 11 and 21 in figure 10.1). Some trace the global chains of particular commodities that are vital to the city. Some focus on consumption practices, some link up with ecological campaigns. Some link particular communities within London to other parts of the world – people from Nigerian groups linking to the Saro-Wiwa project, for example. A way, again, of thinking multiculturalism outwards. Importantly most of them involve support for and active engagement with struggles elsewhere. They are two-way relations. This is not 'responsibility' conceived of as the one-way generosity of the bountiful and powerful (nor, indeed, as Young so clearly argues, does it equate with 'guilt'). It is an engaged attempt to rearticulate relations. A way of encouraging a politics, and even more fundamentally a sensibility, that is outward-looking. A different kind of geographical imagination.

However, were such a politics to be built upon it would throw open again the question of the identity of London and Londoners. It would return the argument once again to the question of the *internal* geographies of place. What is this 'London' (or any other locality)?; who is this 'we' that may be hailed in reference to place? On the one hand, a renegotiation of identity is central to such forms of political organising. Featherstone refers to the reconfiguring of place-based identities and the generative effects of transnational political organising. These more rhyzomorphic, routed and productive practices of solidarity generate forms of equivalences and 'alliances between unlike actors [and] refuse what Spivak has termed "a homogenous internationalism" (Spivak, 1985, p.350). Their activity is productive, continually formed and at times unsettles fixed identities' (2003, p. 406). Such remoulding of identities can occur in numerous ways. If the Caracas–London deal were to go ahead it would shift London's identity (and indeed that of Caracas) both materially and symbolically. In the case of grass-roots alliances, identities could shift through more personal connection maybe, and through processes of political learning about each other's situation and political aims. On the other hand, identities will be renegotiated among Londoners themselves. Even if it is accepted, as it is here, that all members of this cosmopolitan heterogeneity are in some degree 'Londoners', they/we are not all so in the same way. Londoners, as would be the case in any place, are located in radically contrasting and unequal ways in the various processes, both benign and problematical, of today's globalisation. Places are meeting-places of multiple trajectories whose material co-presence has to be negotiated (chapter 3). Any campaigns and strategies, such as the ones above, designed to look outward from place, would add to the complexity of that negotiation. They might indeed underline the conflicting interests within the place; crucially they would highlight the structural connections between inequality at a global level and the inequality within the city. Moreover, regardless of structural position there will be political differences about such issues. The proposal for London–Caracas solidarity provoked immediate hostility from some other parties. (Even Chávez's visit had generated controversy.) 'The leader of London's Tories', reported Muir in *The Guardian* delightfully, 'dismissed the scheme as a "socialist propaganda fest".' The leader of the capital's Liberal Democrats seemed concerned precisely about losing rank within the power-geometries of globalisation: he was reported as saying 'the deal smacked of aid, not trade – "This reduces us to the status of a third-world barter economy"' (Muir, 2006). So there is political contest. Similarly, many of the campaigns just outlined have been initiated at grass-roots level. And although it is enriching of local democracy that there should be such campaigns, it should not be assumed that all such 'bottom-up' organisation will be of the left. To imagine that it will be is a common form of spatial fetishism (see chapter 8) on the left (see also the debate in Held, 2005). However, that very fact of variability is part of the point – that there will be political contest here, within place, as part of the definition, the struggle over the identity, of place.

So there would be conflict. Any of the campaigns and proposals discussed above, and any like them, will be contested. Any 'we' that is constructed here will emerge from conflictual debate – open political debate rather than that sealing-over of conflicting interests orchestrated by the powerful that occurs, for instance, on the issue of 'the London deficit'. This, then, is different from those new localisms that appeal to place as the hearth of some unproblematical collectivity. On the contrary, 'place' here is not taken as given; it is an ongoing product of an agonistic democracy. This is also different from those new localisms of community politics that demand an immersion in local place. On the contrary, here what is required, in order to take responsibility for that place in its wider setting, is what Montesquieu first called for: that 'we must learn to practice a systematic form of disloyalty to our own local civilization if we seek either to understand it or to interact equitably with others formed elsewhere' (Gilroy, 2004, p. 79).

To be wary of certain forms of localism, and certain arguments for a place-based politics, is not to deny their potential *tout court*. Rather it is to require their reformulation. This is a localism turned inside out, and one that has to be struggled over internally. And as such 'place' would seem to have real and, maybe ironically in this age of globalisation, even increasing potential as a locus of political responsibility and an arena for political engagement. It is one base, among many others, for collectivity. It is, for instance, a potential forum for going beyond the politics of the individual. Thus, in relation to the politics of consumption, Barnett argues that 'It is only by being joined into part of wider communities of action, by activists and organisations, that everyday habits of consuming can really be thought of as "ethical", or even political' (in Littler, 2005, p. 150). Moreover, campaigning around place is also different from, though quite possibly allied to, joining a group composed of the already like-minded. It sets up an immediate and relatively accessible *arena* for political contest. It is an arena that extends beyond the individual yet will always pose challenges to any proposed unanimity. For what are at issue are the responsibilities of place, and the configurational politics that the recognition of such responsibilities requires: an understanding of the wider geographies of the relational construction of the identity of place, the political consequences in terms of implication and – also – the hard work entailed in the fact that in each place this identity, these geographies, and these political consequences will be specific. Only with all of this can we address the question that was posed in the Introduction, and which could/should be asked of anywhere: what does this place stand for?

CONCLUDING REFLECTIONS

The rise of global cities has been integral to the assertion and spread of neoliberalism. Within the United Kingdom, and more widely, it is a product of the new social settlement that has gained hegemonic status since the 1980s. The project of class restoration of which neoliberal theory has become the doctrinal armoury was born in cities, cities are the home-base of the new stratum of super-rich, it is from such cities that the Washington Consensus and its descendants are orchestrated. The cities themselves are drawn into competition with each other for position on the global stage. The apparent inevitability of all this merely confirms its intellectual hegemony.

Yet the story is also more complicated. For one thing, global cities are not crucial to this project only as the sites of economic and cultural organisation. They have also become crucial bargaining chips, vital components in the struggle to assert neoliberalism politically. Within urban policy discourse around the world a global-city rhetoric has emerged 'that is used by political and business groups to formulate and/or legitimize neoliberal development' (Olds and Yeung, 2004, p. 500). We have seen this here in the case of London. Within the city it is (a particular formulation of) world city-dom that is assumed to offer the inevitable economic way forward and yet that sits uneasily with that other formulation – in terms of street-level conviviality – of its character. In fact, as we have also seen, not only is this city, as are many cities, above all diverse, but also the reason for its economic health over recent decades (if a city of such inequality can be characterised as healthy) is not by any means only its global nature but rather its promotion of and benefiting from the political project of financialisation, deregulation and mar-ketisation. At inter-regional level too, within the country as a whole, it is this mobilisation of global citydom, buttressed by geographical imaginaries of golden geese and autonomous competing regions, that provides the legitimisation for a regional strategy that in fact enables the continued growth of London/the South-East above all else.

There are other apparent incoherences too. For instance, within each of the spatial realms investigated here (intra-urban, inter-regional, international) it is the very proponents of market forces who, while severely imposing them upon others, do not obey their own rules. Within London itself those at the lower-paid end of the labour market are subject to a competition intensified by inward migration, while those in the boardrooms evade market forces by their mutual awards of increments and pension packages. Within the UK the regions and nations of the North and West are sternly lectured on the need to stand on their own feet, and warned of the dismal long-term impact of relying on state help, while the constant crises of congestion in London/the South-East are attended to by special measures and state aid for new growth areas. Surely a true neoliberal would let the place clog up and allow growth in consequence to be diverted elsewhere. Likewise in the international arena, while poorer countries struggle with the rules of structural adjustment, and are forced to open up their borders, some of the richest countries maintain protectionist barriers. The contradictions are endless: 'The two economic engines that have powered the world through the global recession that set in after 2001 have been the United States and China. The irony is that both have been behaving like Keynesian states in a world supposedly governed by neoliberal rules' (Harvey, 2005, p. 152). And there is of course also the rather different point that the application of neoliberal rules has not been notable for its economic success – the neoliberal period has not been one of high growth (Held, 2004).

The result has been, at all levels, increasing inequality (see, in particular, Duménil and Lévy, 2004), and one of the aims of the sections in this book has been to point to the structural interconnectedness between these inequalities: between those within London and those between the regions of the UK for example; between the difficulty of reproducing London and the constant reproduction, now in sharpened form, of the national North–South divide; between the poverty within London and that in countries and cities around the planet within which the privileged of London hold such a powerful position. Above all, the argument is that an understanding of these geographies of connection is important in the formulation of a political response.

The case of London is, of course, specific. Its particularity lies especially in its pre-eminence in the fields of finance and business services, broadly defined. In that sense it has been crucial to the wider project. As Harvey argues: 'It is worthwhile recalling that one of the conditions that broke up the whole Keynesian post-war Bretton Woods system was the formation of a eurodollar market as US dollars escaped the discipline of its own monetary authorities' (2005, p. 141), and in that key movement London was central. And it has benefited massively since, picking up the threads of old Empire to build a new one through which financial tribute could once again be collected. It is this that is now asserted to be essential to the national economy: 'the main driver', 'the national engine of growth', the golden goose, and all the other persuasive terminologies of the new elite and those who have, in one way or another, been convinced by them. Yet even within the country the case is at the very least ambiguous. Some of the downsides of this model of growth have been explored here: the increase in inequality through the expansion of the stratum of the 'filthy rich' (Mandelson); the intensification of poverty within London; the problems of the social reproduction of the capital; the draining from other regions of professional workers; the wider contribution of the spatial concentration of this stratum and of the escalator effects of London/the South-East to the perpetuation of the national North–South divide; the potential for this aspect of London-world-city, with its attendant inequalities, to conflict with that other claim, to be a city of productive cultural mixity; and finally – but by no means least – its flaunting, as an attitude to be emulated, of outrageously conspicuous consumption and just plain greed. Other economies and societies flourish without such global cities. Scandinavian countries come to mind. Other places are continually deemed more liveable.

Yet if London is specific, as is every place, many of the lessons that can be drawn from its analysis are generalisable. One, which has run throughout this book, is the significance in all this of 'geography' and of the intricate spatialities of class and politics. Taking the analysis of this neoliberal project seriously calls for the reformulation of some classic spatial issues. There is the way, for instance, that neoliberalism, in its specific, British, Blairite manifestation, has been carried through into spatial politics, for instance into the address to 'the regional problem', and how that itself was integral to New Labour's picking up of the baton from Thatcherism to establish at a more grass-roots level, and specifically over a wider geographical canvas, the new hegemonic common sense. This is not a serious regional policy. Unless the national geography of power and inequality, and the economic model on which it rests, is addressed head-on, the elite of London/the South-East will continue to soar away.

Geography is also important in the dynamics of inequality. The intersplicing of national and local inequality has been explored at length here. But it is also more than that. It is now well established, even if not all will listen, that inequality in itself matters; that addressing the problems of poverty means tackling the rich as well as the poor. What has been evident here is how the *geographies* of inequality modulate, and frequently exacerbate, those dynamics. Within place (here, within London) the juxtaposition of need and greed sets up reverberations throughout urban society; the spatial concentration of the elite only further increases their self-absorption and their distance from the rest. Between regions, the existence of uneven development can introduce distortions and rigidities into the national economy. The studied obliviousness to this is yet another way in which the generalised neo-Labour project has been carried through into the realm of the spatial.

Within all this, questions of identity have inevitably become more complex. There is a reciprocity of multiplicity between the identity of place and the identities of multiply placed people. But it is also more than this. It becomes necessary to ask, when speaking of global cities: *whose city* is at issue here? We have seen how some claims to place identity cover over the inequalities within, in order to evade that responsibility (or even the posing of the question of responsibility) that lies within the place itself. It has been shown how the very characterisation of cities as 'global' is a strategy whereby the part stands in for the whole, where the city is defined by its elite and the rest are consigned to invisibility. In both these manoeuvres, cities of the many are effectively claimed by the few. It has been argued that the question of the identity of this place must take account not only of the outside within, the internal hybridity, but also, as it were, of the inside without; that the question 'where does London (or any city) end?' must at least address the issue of those recruited into the dynamics of the urban economy and society by the long lines of connections of all sorts that stretch out to the rest of the country and on around the planet. And this in turn raises questions of unequal interdependence, mutual constitution, and the possibility of thinking of placed identity not as a claim *to* a place but as the acknowledgement of the responsibilities that inhere in *being placed*.

A new social settlement may currently hold sway, but hegemonies are there to be contested, and another of the aims here has been to argue, and to exemplify, that an alternative politics is possible. In his consideration of the possibility of spreading outwards the implications of the demotic cosmopolitanism that exists within the city, Gilroy writes of 'the challenge of being in the same present' (2004, p. 74). This is precisely the challenge of space (Massey, 2005), the challenge of the full recognition of coeval others. One thing it inevitably entails is the acknowledgement of implication, through mutual constitution, in the ongoing production of difference and inequality around the world. This is the same reasoning that would argue against a politics only of aid (or only of 'hospitality' or of 'generosity') on the grounds, in part, that such a formulation occludes the unequal relations in which we are all embedded and through which, again in part, the very need for aid has in the first place been produced. Rather, what are at issue are the responsibilities of place. These may concern the politics within the city, the question of the city within the country, or the question of the city in the wider world. But in any case, this is a 'local' politics that asserts and actively politicises both the fact of multiplicity within and the essential openness of place to the beyond.

NOTES

Introduction: 'the future of our world'?

1 Indeed, presidents and prime ministers were on that day
   holed up at Gleneagles in Scotland for a summit. They
   were heavily guarded against attack, from terrorists and
   from demonstrators. On the previous Saturday a Live8
   concert in London had attracted thousands. On 15 February 2003, people filled the streets of London in protest
   against the imminent invasion of Iraq. Note, for later,
   that the mayor is popularly known as 'Ken'.

2 In more detail, the figures were: On satisfaction: 44 per
   cent fairly satisfied and 27 per cent very satisfied, with
   Asians most likely to say they were very satisfied, and
   Asians and Black groups more positive than other groups
   about living in London. On identification: 42 per cent
   agreed strongly with the statement 'London is a place I
   identify with', and 36 per cent tended to agree; in this
   case Black groups were most likely to agree that they
   identified with London. On spontaneous mentions in

response to the question 'What, if anything, would you say are the two or three things that make you proud of London?', 21 per cent mentioned history/heritage, 16 per cent cultural diversity (the second highest response) and 8 per cent the fact that the city is cosmopolitan. Again there was differentiation between groups: 26 per cent of Whites mentioned history/heritage and 27 per cent of 'Other Ethnic' background mentioned cultural diversity. The research was conducted for the Commission on London Governance.

3 This is partly addressed in what follows, but it should be noted that, as well as ongoing racism and intercultural conflict, there was also an increase after the bombings in attacks on Muslims. For a general report see GLA (2005d).

4 'Negotiation' is meant here in its widest sense. It may be the quotidian business of sharing space with barely a mutual recognition; it may be significantly interactive; it may result from the imposition of hegemonic norms. The point is that it is an ongoing process. The settlements that are 'places' are socio-historical and geographically specific, provisional, outcomes.

5 'Rhetoric' because of course in practice 'the market' is considerably aided, organised and intervened in.

6 Based on the annual executive-pay survey conducted for *The Guardian*. Total remuneration is calculated as including bonuses and gains from long-term incentive plans. The survey also pointed out that, in a period which for most people is one of pensions crisis, 'More than £880m has been set aside by the UK's top 100 companies to finance the pensions of their directors' (Finch and Treanor, 2005).

7 'The global' in this debate is usually taken as referring to capitalist forces which are assumed, as in much left discourse, to have primarily negative repercussions on

local places. This is in no way meant to imply a general proposition that 'the global' is 'bad' and the local 'good'. Such an imaginary is a form of spatial fetishism, substituting spatial form for real political questions (Massey, 2005). The political position adopted in the argument here argues not for a replacement of the global by the local, but for an alternative form of globalisation. See especially Part III.

8 For an extended discussion of coevalness, and its relation to implicit conceptualisations of space and time (political cosmologies), see Fabian (1983).

Chapter 1 Capital delight

1 The literature on such cities is vast, from the prescient writing of Hall (1966), through Hymer (1972), Friedmann and Wolff (1982) and, especially, Sassen (1991). For one review, see Beaverstock, Smith and Taylor (1999) and, in general, see the work of the Globalization and World Cities Group and Network, based in the Geography Department at Loughbrough University – www.lboro.ac.uk/gawc.

2 The detailed picture is more complex than this. Beaverstock, Smith and Taylor (1999) identify three world regions in which world-city formation (on their producer-service orientated criteria) has proceeded: Northern America, Western Europe and Pacific Asia.

3 Gordon is referring in his study to the mobilisation of representations of the city by London's political government, but as we shall see later the same argument can be made in relation to other elements in the socio-economic complexity that is London, perhaps most particularly elements of capital and specifically the City.

4 The distinctiveness from other global cities resides in the importance to London of third-party business 'i.e.

business undertaken purely for overseas clients, rather than wholly/partly on behalf of a UK-based business. In this respect, London is clearly much more of a global city than Tokyo, and probably also than New York, both of which tend to act more as the international agents of Japanese/US-based corporations' (Gordon, 2004, p. 5).

5 It is common economic practice to divide an area's economy into 'basic' economic activities (those which by exporting beyond the area bring in income to the area) and 'service' activities, which are marketed internally to an area and could thus be seen as part of maintenance or simple reproduction of the area. It is a useful but problematical distinction. The problems are methodological (the difficulty of distinguishing between such activities), conceptual (at world level all activities must be 'service activities') and empirical (it tends to lead to an underestimation of the significance of those activities deemed to be only 'maintenance') – and this last can have political effects – see below. See also Perrons (2001).

6 Further details of this last part of the argument are given in Buck et al., 2002 and Gordon, 2002.

7 At the micro-level, the socio-cultural construction and embeddedness of economic processes is increasingly recognised (see, for instance, Granovetter and Swedberg, 1992). Since then, a whole stream of work has emerged in economic geography that elaborates and emphasises these 'non-economic' 'factors' in regional growth and decline. Unfortunately it, too, with its unquestioning retailing of notions of trust, reciprocity, institutional characteristics, and so forth, its lack of interrogation of the social terms of regional 'success', and its apparent belief in and commitment to the replication of strategies between regions, can sometimes fail to take on board the

wider economic and political structures within which these processes are set, and thereby not only depoliticises them but effectively endorses the current technicist view of economic growth. For a critique of this kind of work, see Hadjimichalis (2006).

8 Though Crouch's (2005) argument against typologies of this kind, and in favour of a 'recombinant' approach more amenable to the recognition of specificity, is important.

9 This, too, is a characteristic of many cities striving to become global. For an excellent exploration of (among other things) the role of the real-estate industry in Vancouver's bid for world citydom, see Mitchell (2004).

Chapter 2 'A successful city, but . . . '

1 Much of the information in this paragraph is drawn from *London divided*, a meticulous and pioneering study produced by the Greater London Authority (2002) (which itself draws on a wide range of sources) and followed up by *Tackling poverty* (2003). There is no doubt that this issue is of central and active concern to the GLA. Page numbers in this paragraph refer to *London divided*.

2 See Buck et al., 2002, chapter 4, for a meticulous discussion of this and other related hypotheses, and also for an investigation of the details of movements on various dimensions of inequality. The global cities hypothesis is most associated with Sassen (1991), and has been criticised for instance by Hamnett (2003).

3 The observer is Krugman (2002).

4 Nonetheless, the document goes on to detail a careful and wide-ranging set of policies.

5 This, too, is a problem that affects other 'global' cities – see, for instance, Smith (2002) on New York.

6 The concern here has been with London, but many of these tensions can also be found in the South-East region that surrounds the metropolitan area itself. This too is a region of rapid growth, taking much of its impetus from its locational relation to London but developing now its own relations of interconnectivity. While its policy aim is to be among the growth elite of Europe, this does bring difficulties in its train (Cochrane, 2006). As Cochrane argues, the region is seen as being essential to national growth (see also Part II here) and in policy terms is aware of the fact, that in the 1980s, overheating and labour constraints in the region were seen as having held back national growth (another way of interpreting this is that national growth should have been less regionally concentrated – another way in which the geographies of inequality have their own autonomous effects. See Part II). Yet again in the twenty-first century there are problems. Prime among them is the squeeze on affordable housing for local people: Cochrane cites from a MORI survey: 'We are not really a proper region. . . . There is no community of interests. The only thing that links us is affordable housing' (SEERA, 2004; cited in Cochrane, 2006, p. 229). Moreover, as already indicated, the South-East, like London, is plagued with inequality, a fact that, again as in the case of London, is ritually stressed in the bidding war of competitive regionalism (see Part II). It is once again, moreover, an inequality that deepens the experience of poverty in the region (see GOSE, 2002, pp. 40–1, cited in Cochrane, 2006, p. 232). Finally, and this time rather differently from metropolitan London, there are in the South-East severe strains and tensions between the desire for growth on the one hand and environmental concerns on the other (Cochrane, 2006, p. 234).

Chapter 3 Imagining the city

1 The spatial dynamics of this whole period, including the intersection of the economic shifts with the political geography and the changing class structure, are examined in detail in Massey ([1984] 1995).

2 The regional policy of this period played an enormously complex role in the national economy, and its role and effectivity changed over the period. It was also hotly contested on political grounds, within different regions, around issues of gender, and over its relation to the already established intentions of those elements of capital that took advantage of it. The critiques now so casually thrown around by 'advisers' to New Labour are blissfully unaware of any of this complexity.

3 The GLC's (1985) strategy in this situation, however, was to focus on those office workers who were less privileged – to encourage unionisation, improve security and conditions of labour, and work against sex and race discrimination; see its chapter 14.

4 The analysis here differs radically from that of Harvey, who is strangely sympathetic to Blair, arguing that: 'Perhaps the greatest testimony to [Reagan and Thatcher's] success lies in the fact that Clinton and Blair found themselves in a situation where their room for manoeuvre was so limited that they could not help but sustain the process of restoration of class power even against their own better instincts. . . . [They] could do little more than continue the good work of neoliberalization, whether they liked it or not' (2005, pp. 62–3). I think this overestimates Thatcher's achievement of hegemony and underestimates the enthusiastic complicity of Blair. 1997, when Blair was elected, was still a moment of what we called at the time 'historic opportunity'. As Hall has written

The Labour election victory in 1997 took place at a moment of great political opportunity. Thatcherism had been decisively rejected by the electorate. But 18 years of Thatcherite rule had radically altered the social, economic and political terrain of British society. There was therefore a fundamental choice of directions for the incoming government. One was to offer an alternative radical strategy to Thatcherism . . . The other choice was, of course, to adapt to Thatcherite/neo-liberal terrain. (Hall, 2003, pp. 10–11)

It is important to stress this because it emphasises the lack of inevitability, and the consequent possibility of politics. The 'feeling in the air' in May 1997 was of hope and the openness of the future. And, as Žižek has written, 'Even when the change is not substantial but a mere semblance of a new beginning, the very fact that a situation is perceived by the majority of the population as a "new beginning" opens up the space for important ideological and political rearticulations – . . . the fundamental lesson of the dialectic of ideology is that appearances *do* matter' (1997, p. 48). It was indeed at this juncture (in fact 1995), and in recognition of the choices ahead, that Michael Rustin, Stuart Hall and I founded *Soundings: a journal of politics and culture* (published by Lawrence & Wishart).

5 When I gave evidence at the Scrutiny Committee I had stupidly mentioned in my mini-biog my membership of GLEB – stupidly because, rather than being treated as evidence of London experience, it led to Tory attempts to undermine what I had to say simply by (tabloid-type, and incorrect) vilifying references to that body.

6 These strategies continue to roll out, and partly for that reason the focus here is on the plans rather than on their

implementation. Some of the big strategies that are well under way – the Freedom Pass for the over sixties, for instance – clearly reflect the aim of redistribution that runs through the documents. At recent meetings of some local groups concern has been expressed about perceived lack, or slowness, of implementation. The argument which follows also implies that even full implementation might not achieve the degree of redistribution desired. Nonetheless the degree of emphasis on inequality as such distinguishes this approach from the usual formulations of New Labour.

7 Mitchell (1993, 2004) has written of how multiculturalism has been, in the case of Vancouver, 'politically appropriated by individuals and institutions to facilitate international investment and capitalist development' (1993, p. 265), and it is interesting that, again, the specific connection was with Chinese capital. It is important to recognise both this kind of possibility specifically and, more generally, the potential for 'multiculturalism' to play a role in hegemonic economic, as well as cultural, processes. However, the London case is more mixed than, and different from, that of Vancouver. As Mitchell argues, each case will be specific.

8 Massey (2001) presents some possible elements for a different approach to economic strategy in London. Livingstone himself addresses these issues, of inequality and of London's prioritising of finance and business services, in an interview in *Soundings: a journal of politics and culture*, 36 (summer 2007).

Chapter 4 The golden goose?

1 The dispute which is referred to here revolves primarily around the degree to which intra-regional inequality overrides inter-regional inequality – in other words, are the disparities between localities within regions so great

that they obliterate a wider 'North–South' pattern? Tony Blair tried out this argument on occasions, but it has never achieved much serious purchase. It is, of course, an argument mobilised by some within London (see chapter 6).

2 It should be stressed that, just as in the case of neoliberalism more generally, and as in the case of the installation of mayors, there was here no automatic transmission belt of government desires. Regions and local councils responded in a variety of ways. The power resources of the centre, as always, had to be negotiated, were relational, and were always situated (Allen, 2003).

3 The rubbishing of the regional policy of previous decades is usually done without any evidence, any knowledge of history and any recognition that, anyway, times (and geographies, and the structuring of geographies) have changed and so therefore must the forms of intervention. It also exemplifies New Labour's classic rhetorical strategy of dismissing alternatives to their own view as 'old fashioned', thus evading any specifically political (or even in this case seriously technical) debate. In the 1970s and 1980s I was a critic of regional policy (see, for example, Massey, 1979) but I do not at all subscribe to the uninformed sneering that is common today in New Labour.

4 This, too, is a widely used discursive strategy. See, for other examples, Mitchell (2004) on Vancouver.

Chapter 5   An alternative regional geography

1 Elliott is here drawing on Dorling and Thomas (2004), who argue that there is a net movement of population from north to south within the UK. The London School of Economics study strongly disputes this (indeed it disputes a number of Dorling and Thomas's conclusions, which it describes as 'myths' (LSE, 2004, p. 11)). However, London and the South also receive

population from outside the UK, and their population is certainly increasing and set to increase further. Acceptance of this, indeed, is the whole basis of the new directions taken by the *London plan* under the new GLA.

2 Some notable exceptions include the Institute for Public Policy Research, which has established an office in the North-East of England.

3 Oxford Economic Forecasting advance a further proposition: that 'a "brain drain" to London from the other parts of the UK may be a way of avoiding a "brain drain" out of the UK altogether' (OEF, 2004, p. 54). It 'may be', indeed, but no direct evidence is advanced that it actually *is*. Nonetheless, in spite of the tenuousness of this suggestion, it is seized upon in the foreword to the report, which states that 'one of the conclusions on education and skills captures the key message of the report as a whole: "A 'brain drain' to London from other parts of the UK may be a way of avoiding a 'brain drain' out of the UK".' If this is evidence of anything, it is evidence of the sensitivity of the issue. It is a mobilisation of a more general 'global city' thesis.

Chapter 6 Who owes whom?

1 It should also be noted that the failure over many years to recalibrate council tax (which is based on property values) to reflect regionally differentiated rises in property values is both another effective transfer from people in the North to people in London and the South-East and is another example of national government pusillanimity in the face of voters in the latter places.

2 The structure of the budget for the GLA is very different from that of the GLC and seems almost designed to work against intra-local redistribution. 60 per cent of the GLA budget comes from the national Treasury. In the days of the GLC, by contrast, 60 per cent came from

business and thereby in itself worked towards redistribution (Livingstone, pers. comm.). In an interview in *Soundings: a journal of politics and culture*, 36 (summer 2007), Livingstone discusses in more detail this difficulty of effecting local redistribution.

Chapter 7 Reworking the geographies of allegiance

1 Though it has to be said that it is very difficult to get journalists away from this easy headline. One television interview I did, in which I stressed again and again that this was *not* anti-London, had this point totally edited out, and was introduced as: an anti-London argument!

Chapter 8 Grounding the global

1 For an extended critique of this imaginary, see also the work of Sparke, especially Sparke (2005). There is also a much fuller discussion in Massey (2005).

2 For further elaboration of these projects, see Cameron and Gibson (2001), on the Latrobe Valley, and Community Economies Collective (2001), on the Pioneer Valley. These projects grow out of Gibson-Graham's earlier book *The end of capitalism (as we knew it): a feminist critique of political economy* (1996).

Chapter 9 Identity, place, responsibility

1 Robinson (1999) makes this point well, and presents a telling case for 'globalising care'.

2 This case for a greater aspect of 'outward-lookingness' more generally is central to the argument of Massey (2005).

3 Just to be clear, there is no underlying implication here that the temporally distant (the historical past) did not have its spatial distantiations too. In the case of Britain's Empire this intersplicing of temporal and spatial distance is at the very heart of the matter.

4 It's also worth pointing out that subsidiarity was widely marketed as though it were a technical matter – a purely rational organisation of activities and powers. (Obviously, the discourse implied, bus stops are a local issue while globalisation is one for bigger entities to deal with.) It is not so simple; and it is definitively political rather than merely technical.

5 The conference was 'Lost in space: topologies, geographies, ecologies', organised by Stour Valley Arts in Canterbury in February 2006. It should be said that the question was posed constructively, enquiringly, and not at all defiantly!

Chapter 10 A politics of place beyond place

1 The Medact report (Mensah, Mackintosh and Henry, 2005) addresses head-on the conflicting political principles at issue here.

2 And these inequalities are multiple: in wages, in facilities, in levels of care, in numbers of doctors and nurses per unit of population.

3 There are, however, recent indications that this contradiction too will be firmly addressed. In response to questioning on this, Livingstone replied: 'When we drafted the London Plan in 2002, we were nowhere near getting the alarming information that we are today. We have to address it. We are now preparing amendments to the plan against any further runway capacity in the southeast' (see Vidal, 2006).

4 London is also widely seen as a tax haven at a more personal level, for the seriously rich (Lansley, 2006, especially chapter 9). It is also a centre for instruments such as hedge funds: 'It is no coincidence that derivative trading and the investment activities of hedge funds, two of the areas causing most concern in debates on the global

financial architecture, are both closely associated with the offshore system' (Oxfam GB, 2000, p. 18).

5 http://risingtide.org.uk and www.artnotoil.org.uk.

6 www.platformlondon.org.

7 For 'Unravelling the carbon web', see www.carbonweb.org; the newsletter is at www.carbonweb.org/documents/utcw_Newsletters02.pdf.

8 www.remembersarowiwa.com.

REFERENCES

Ackroyd, P., 2000, *London: the biography*, London: Chatto & Windus.

Adams, J., Robinson, P., and Vigor, A., 2003, *A new regional policy for the UK*, London: IPPR.

Adonis, A., and Pollard, S., 1998, *A class act: the myth of Britain's classless society*, London: Penguin.

Allen, J., 2003, *Lost geographies of power*, Oxford: Blackwell.

Allen, J., and Henry, N., 1997, 'Ulrich Beck's risk society at work: labour and employment in the contract services industry', *Transactions of the Institute of British Geographers*, 22, pp. 180–96.

Allen, J., Massey, D., and Cochrane, A., 1998, *Rethinking the region*, London: Routledge.

Amin, A., and Graham, S., 1997, 'The ordinary city', *Transactions of the Institute of British Geographers*, 22, pp. 411–29.

Amin, A., Massey, D., and Thrift, N., 2000, *Cities for the many not the few*, Bristol: Policy Press.

Amin, A., Massey, D., and Thrift, N., 2003, *Decentering the nation: a radical approach to regional inequality*, London: Catalyst.

Armitage, J., 2003, 'US consultants accused in row over fat cat pay', *Evening Standard*, 23 May, p. 41.

Banks, R., and Scanlon R., 2000, 'Major economic trends in the 1980s and 1990s: London', in *The London–New York study*, London: City of London Corporation.

Batchelor, C., and Larsen, P. T., 2006, 'London's financial centre basks in its global appeal', *Financial Times*, 27 March, p. 1.

Beaverstock, J. V., Smith, R. G., and Taylor P. J., 1999, 'A roster of world cities', *Cities*, 16/6, pp. 445–58.

Bender, T., 1999, 'Intellectuals, cities, and citizenship in the United States: the 1890s and 1990s', in J. Holston (ed.), *Cities and citizenship*, Durham, NC; Duke University Press, pp. 21–41.

Biles, A., 2001, 'Regions hit by London "brain drain"', *Regeneration and Renewal*, 6 July, p. 5.

Boal, I., Clark, T. J., Matthews, J., and Watts, M. [Retort], 2005, *Afflicted powers*, London: Verso.

Bosanquet, N., Cumming, S., and Haldenby, A., 2006, 'Whitehall's last colonies: breaking the cycle of collectivisation in the UK regions', *Reform*, July.

Brenner, N., and Theodore, N., 2002, 'Preface to special issue: from the new localism to the spaces of neoliberalism', *Antipode*, 34/3, pp. 341–8.

Buck, N., Gordon, I., Hall, P., Harloe, M., and Kleinman, M., 2002, *Working capital: life and labour in contemporary London*, London: Routledge.

Cameron, J., and Gibson, K., 2001, *Shifting focus: pathways to community and economic development – a resource kit*, Latrobe City Council and Monash University, Victoria.

Checkland, S., 1976, *The upas tree: Glasgow 1875–1975: a*

study in growth and contraction, Glasgow: Glasgow University Press.

Christensen, J., 2007, 'Secret world of offshore banking', in S. Hiatt (ed.), *A game as old as empire*, San Francisco: Berrett-Koehler, pp. 41–67.

Clark, N., 2006, 'Blinded by the cold war', *The Guardian*, 29 August.

Cochrane, A., 2006, 'Looking for the South East', in I. Hardill, P. Benneworth, M. Baker and L. Budd (eds), *The rise of the English regions?*, London: Routledge, pp. 222–44.

Cochrane, A., 2007, *Understanding urban policy: a critical approach*, Oxford: Blackwell.

Cohen, N., 2004, 'Without prejudice: get ready for Kengrad', *The Observer*, 22 February.

Cohen, R., 1981, 'The new international division of labour: multinational corporations and the urban hierarchy', in M. Dear and A. J. Scott (eds), *Urbanization and urban planning in capitalist society*, London: Methuen.

Commission on Urban Life and Faith, 2006, *Faithful cities*, Peterborough: Methodist Publishing House and Church House Publishing.

Community Economics Collective, 2001, 'Imagining and enacting noncapitalist futures', *Socialist Review*, 28/3–4, pp. 93–135.

Compass, 2006, *The good society: Compass programme for renewal*, ed. J. Rutherford and H. Shah, London: Compass in association with Lawrence & Wishart.

Coronil, F., 2000, 'Towards a critique of globalcentrism: speculations on capitalism's nature', *Public Culture*, 12/2, pp. 351–74.

Corporation of London, 1986, *City of London local plan*, London: Department of Architecture and Planning, Corporation of London.

Cox, R., and Watt, P., 2002, 'Globalization, polarization and the informal sector: the case of domestic workers in London', *Area*, 34/1, pp. 39–47.

Crouch, C., 2005, 'Models of capitalism', *New Political Economy*, 10/4, pp. 439–56.

CSFI (Centre for the Study of Financial Innovation), 2003, *Sizing up the City: London's ranking as a financial centre*, London: Corporation of London.

Daneshkhu, S., and Giles, C., 2006, 'City becomes undeniable engine of growth', *Financial Times*, 27 March, p. 2.

Daniels, P. W., 1991, *Services and metropolitan development: international perspectives*, London: Routledge.

Davis, M., 2004, 'Planet of slums', *New Left Review*, no. 26, pp. 5–34.

Davis, M., 2006, *Planet of slums*, London: Verso.

Denham, J., 2004, 'The case for a "New Labour" third term', *Renewal*, 12/4, pp. 72–8.

Derrida, J., 1997, *Politics of friendship*, London: Verso.

Derrida, J., 2001, 'On cosmopolitanism', in Derrida, *On cosmopolitanism and forgiveness*, trans. M. Dooley and M. Hughes, London: Routledge [first pubd 1997 as *Cosmopolites de tous les pays, encore un effort*, Paris: Editions Galilée].

Devine, P., 2006, 'The 1970s and after: the political economy of inflation and the crisis of social democracy', *Soundings: a journal of politics and culture*, no. 32, spring, pp. 146–61.

Dickson, M., 2006, 'London: capital gain', *Financial Times*, 27 March [report on *The new City*].

Diprose, R., 2002, *Corporeal generosity: on giving with Nietzsche, Merleau-Ponty, and Levinas*, Albany: State University of New York Press.

Dorling, D., and Thomas, B., 2004, *People and places: a 2001 Census atlas of the UK*, Bristol: Policy Press.

Douglass, M., 1998, 'World city formation on the Asia Pacific rim: poverty, "everyday" forms of civil society and environmental management', in M. Douglass and J. Friedmann (eds), *Cities for citizens*, Chichester: Wiley, pp. 107–38.

Duménil, G., and Lévy, D., 2004, *Capital resurgent: roots of the neoliberal revolution*, trans. D. Jeffers, Cambridge, MA: Harvard University Press.

Dunford, M., 2003, 'Theorizing regional economic performance and the changing territorial division of labour', *Regional Studies*, 37/8, pp. 839–54.

Dunford, M., 2005, 'Old Europe, new Europe and the USA: comparative economic performance, inequality and the market-led models of development', *European Urban and Regional Studies*, 12/2, pp. 151–78.

Edwards, M., 2002, 'Wealth creation and poverty creation: global–local interactions in the economy of London', *City*, 6/1, pp. 25–42.

Elliott, L., 2004, 'The United Kingdom of London', *The Guardian*, 5 July, p. 23.

Elliott, L., 2005, 'UN spells out the stark choice: do more for world's poor or face disaster', *The Guardian*, 8 September, p. 17.

Elliott, L., and Moore, C., 2005, 'Migrants hold down inflation says governor', *The Guardian*, 14 June, p. 15.

Escobar, A., 2001, 'Culture sits in places: reflections on globalism and subaltern strategies of localization', *Political Geography*, 20, pp. 139–74.

European Commission, 1999, *The European spatial development perspective*, Luxembourg: Office for Official Publications of the European Communities.

Fabian, J., 1983, *Time and the other: how anthropology makes its object*, New York: Columbia University Press.

Fainstein, S., and Harloe, M., 2000, 'Ups and downs in the global city: London and New York at the millennium', in

G. Bridge and S. Watson (eds), *A companion to the city*, Oxford: Blackwell.

Featherstone, D. J., 2001, 'Spatiality, political identities and the environmentalism of the poor', PhD thesis, Milton Keynes: Open University.

Featherstone, D. J., 2003, 'Spatialities of transnational resistance to globalisation: the maps of grievance of the Inter-Continental Caravan', *Transactions of the Institute of British Geographers*, 28/4, pp. 404–21.

Featherstone, D. J., 2005, 'Towards the relational construction of militant particularisms: or why the geographies of past struggles matter for resistance to neo-liberal globalisation', *Antipode*, 37/2, pp. 250–71.

Finch, J., and Treanor, J., 2005, 'Chief executives' pay rises to £2.5m average', *The Guardian* (international edn), 4 August, p. 1.

Flynn, D., 2005, 'New borders, new management: the dilemmas of modern immigration policies', *Ethnic and Racial Studies*, 28/3, pp. 463–90.

Foucault, M., 1985, *The history of sexuality*, vol. 2: *The uses of pleasure*, New York: Pantheon.

Freedland, J., 2005a, 'It may be beyond passé – but we'll have to do something about the rich', *The Guardian*, 23 November, p. 27.

Freedland, J., 2005b, 'Tread more carefully', *The Guardian*, 27 July.

Freeman, C., 2001, 'Mayor's funding plea "will divide Britain"', *Evening Standard*, 26 June, p. 16.

Friedmann, J., 1986, 'The world city hypothesis', *Development and Change*, 17/1, pp. 69–84.

Friedmann, J., and Wolff, G., 1982, 'World city formation: an agenda for research and action', *International Journal of Urban and Regional Research*, 3, pp. 309–44.

Gamble, A., 1981, *Britain in decline: economic policy, political strategy and the British state*, Basingstoke: Macmillan.

Gatens, M., and Lloyd, G., 1999, *Collective imaginings: Spinoza, past and present*, London: Routledge.

Gibson-Graham, J. K., 1996, *The end of capitalism (as we knew it): a feminist critique of political economy*, Oxford: Blackwell.

Gibson-Graham, J. K, 2003, 'An ethics of the local', *Rethinking Marxism*, 15/1, pp. 49–74.

Gilroy, P., 2004, *After Empire: melancholia or convivial culture?*, London: Routledge.

GLA (Greater London Authority), 2001, *Working families tax credit briefing*, London: GLA.

GLA (Greater London Authority), 2002, *London divided: income inequality and poverty in the capital*, London: GLA.

GLA (Greater London Authority), 2003, *Tackling poverty in London: consultation paper*, London: GLA.

GLA (Greater London Authority), 2004a, *The case for London: London's loss is no-one's gain*, London: GLA [the mayor of London's submission to the Spending Review 2004].

GLA (Greater London Authority), 2004b, *The London plan: spatial development strategy for Greater London*, London: GLA.

GLA (Greater London Authority), 2005a, Press release, 7 July, www.london.gov.uk/mayor/mayor_statement_070705.jsp.

GLA (Greater London Authority), 2005b, Press release, 8 July: 'Commissioner and mayor hold press conference', www.london.gov.uk/news/2005/bombing-statement-080705.jsp.

GLA (Greater London Authority), 2005c, *A fairer London: the living wage in London*, London: GLA.

GLA (Greater London Authority), 2005d, *Race equality scheme 2005–2008, consultation draft: summary*, London: GLA.

GLA (Greater London Authority), 2006, *A fairer London: the living wage in London*, London: GLA.

GLC (Greater London Council), 1985, *The London industrial strategy*, London: GLC.

Godfrey, B., and Zhou, Y., 1999, 'Ranking world cities: multinational corporations and the global city hierarchy', *Urban Geography*, 20/3, pp. 268–81.

Gordon, I., 2002, 'Global cities, internationalisation and urban systems', in P. McCann (ed.), *Industrial location economics*, Cheltenham: Edward Elgar.

Gordon, I., 2004, 'Capital needs, capital growth and global city rhetoric in Mayor Livingstone's London Plan', *GaWC Research Bulletin*, 145, http://www.lboro.ac.uk/gawc/rb/rb145.html [accessed 21 September 2005].

GOSE, 2002, *South East region social inclusion statement*, Guildford: Government Office of the South East.

Granovetter, M. S., and Swedberg, R. (eds), 1992, *The sociology of economic life*, Boulder, CO: Westview Press.

Gregory, D., 2004, *The colonial present*, Oxford: Blackwell.

The Guardian, 2003, 'Unbalanced Britain: the gap between north and south widens', 6 February.

The Guardian, 2005, *London: the world in one city: a special celebration of the most cosmopolitan place on earth*, G2 special, 21 January.

Hadjimichalis, C., 2006, 'Non-economic factors in economic geography and in "New Regionalism": a sympathetic critique', *International Journal of Urban and Regional Research*, 30/3, pp. 690–704.

Hall, P., 1966, *The world cities*, London: Heinemann.

Hall, S., 2000, 'Conclusion: the multi-cultural question', in B. Hesse (ed.), *Un/Settled multiculturalisms*, London: Zed Books, pp. 209–41.

Hall, S., 2003, 'New Labour's double-shuffle', *Soundings: a journal of politics and culture*, no. 24, autumn, pp. 10–24.

Hamnett, C., 1989, 'The political geography of housing in contemporary Britain', in J. Mohan (ed.), *The political geography of contemporary Britain*, Basingstoke: Macmillan, pp. 208–33.

Hamnett, C., 2003, *Unequal city: London in the global arena*, London: Routledge.

Hamnett, C., and Randolph, W., 1982, 'How far will London's population fall?: A commentary on the 1981 Census', *London Journal*, 8/1, pp. 96–100.

Harding, A., Marvin, S., and Robson, B., 2006, *A framework for city-regions, working paper 4: The role of city-regions in regional economic development policy*, London: Office of the Deputy Prime Minister.

Harvey, D., 1989, 'From managerialism to entrepreneurialism: the transformation of urban governance in late capitalism', *Geografiska Annaler*, 71(B), pp. 3–17.

Harvey, D., 2005, *A brief history of neoliberalism*, Oxford: Oxford University Press.

Held, D., 2004, *Global covenant: the social democratic alternative to the Washington Consensus*, Cambridge: Polity.

Held, D., 2005, *Debating globalization*, Cambridge: Polity.

Hesse, B., 2000, 'Introduction: Un/Settled multiculturalisms', in Hesse (ed.), *Un/Settled multiculturalisms*, London: Zed Books, pp. 1–30.

Hill, H., 2003, *The London deficit – a business perspective: an investigation into London's contribution and support*, London: London Chamber of Commerce and Industry.

HM Treasury/DTI (Department of Trade and Industry), 2001, *Productivity in the UK: 3 – The regional dimension*, London: HM Treasury.

Hudson, R., 2006a, 'Regions and regional uneven development forever? Some reflective comments upon theory and practice', paper presented to the annual conference of the Association of American Geographers.

Hudson, R., 2006b, 'From knowledge-based economy to . . . knowledge-based economy? Reflections on changes in the economy and development policies in the north east of England', paper presented to the annual conference of the Association of American Geographers.

Hudson R., and Williams, A. M., 1995, *Divided Britain*, 2nd edn, Chichester: John Wiley.

Hutton, W., 2002, *The world we're in*, London: Little, Brown.

Hymer, S., 1972, 'The multinational corporation and the law of uneven development', in J. N. Bhagwati (ed.), *Economics and world order from the 1970s to the 1990s*, London: Collier-Macmillan, pp. 113–40.

Institute for Fiscal Studies, 2006, *Poverty and inequality in Britain: 2006*, London: IFS.

Jackson, B., and Segal, P., 2004, *Why inequality matters*, London: Catalyst.

Jacobs, J., 1996, *Edge of empire: postcolonialism and the city*, London: Routledge.

Jessop, B., 1979, 'The transformation of the state in post-war Britain', in R. Scase (ed.), *The state in Western Europe*, London: Croom Helm.

Johnson, A., 2002, 'Building a more balanced UK plc', *Regeneration and Renewal*, 13 December, p. 14.

Keith, M., and Cross, M., 1993, 'Racism and the postmodern city', in M. Cross and M. Keith (eds), *Racism, the city and the state*, London: Routledge, pp. 1–31.

Kettle, M., 2006, 'When democracy lost its grip on the City of London', *The Guardian*, 21 October.

King, A. D., 1990, *Global cities: post-imperialism and the internationalization of London*, London; Routledge.

King, A. D., 2000, 'Postcolonialism, representation and the city', in S. Watson and G. Bridge (eds), *A companion to the city*, Oxford: Blackwell, pp. 261–9.

Knox, P., 1993, 'Capital, material culture, and socio-spatial differentiation', in Knox (ed.), *The restless urban landscape*, Englewood Cliffs, NJ: Prentice-Hall, pp. 1–34.

Krugman, P., 2002, 'For richer', *New York Times*, 20 October.

Kundnani, H., 2006, 'Rich get even richer in third world', *The Guardian*, 21 June, p. 27.

Lansley, S., 2006, *Rich Britain: the rise and rise of the new super-wealthy*, London: Politico's.

Larner, W., 2003, 'Neoliberalism?', *Environment and Planning D: Society and Space*, 21, pp. 509–12.

Larsen, P. T., 2006, 'London: why action may be needed to maintain competitive advantage', *Financial Times*, 27 March, p. 4 [report on *The new City*].

Lawson, N., 2006, 'This nation of shoppers needs to talk about class', *The Guardian*, 19 April.

Layard, R., 2005, *Happiness: lessons from a new science*, London: Allen Lane.

LDA (London Development Agency), 2005, *Better food for London: the mayor's draft food strategy*, London: LDA.

Littler, J., 2005, 'Consumers – agents of change?': discussion with Clive Barnett and Kate Soper, *Soundings: a journal of politics and culture*, no. 31, autumn, pp. 147–60.

Livingstone, K., 2005, 'Three ways to make us all safer', *The Guardian*, 4 August.

Livingstone, K., 2006, 'A city for the Asian century', *The Guardian*, 7 April, p. 35.

Local Futures Group, 2003, *A regional perspective on the knowledge economy in Great Britain*, London: Department of Trade and Industry.

Local Futures Group, 2006, *State of the nation 2006: the geography of well-being in Britain*, London: Local Futures.

London Chamber of Commerce and Industry, 2004, *Policy guide*, London: LCCI.

London Voluntary Service Council, 2002, *Response to 'The draft London plan': draft spatial development strategy for Greater London*, September, London: London Voluntary Service Council.

Loney, D., 2001, 'Knowledge-based economy "is creating a more divided Britain"', *Regeneration and Renewal*, 24 May, p. 5.

LSE (London School of Economics), 2004, *London's place in the UK economy*, London: Corporation of London.

McIvor, M., 2005, 'New Labour, neo-liberalism and social democracy', *Soundings: a journal of politics and culture*, no. 31, autumn, pp. 78–87.

Mackintosh, M., 2007, 'International migration and extreme health inequality: robust arguments and institutions for international redistribution in health care', in G. Mooney and D. McIntyre (eds), *The economics of equity in health and health care: future prospects*, Cambridge: Cambridge University Press.

Mackintosh, M., and Wainwright, H. (eds), 1987, *A taste of power: the politics of local economics*, London: Verso.

Mackintosh, M., Mensah, K., Henry, L., and Rowson, M., 2006, 'Aid, restitution and international fiscal redistribution in health care: implications of health professionals' migration', *Journal of International Development*, 18, pp. 757–70.

Marks, M., 2006, 'A protected, cosy old club', *The Guardian*, 27 October, p. 33.

Martin, R., 1999, 'The new "geographical turn" in economics: some critical reflections', *Cambridge Journal of Economics*, 23, pp. 65–91.

Mason, C. M., and Harrison, R. T., 1999, 'Financing entrepreneurship: venture capital and regional development', in R. Martin (ed.), *Money and the space economy*, Chichester: John Wiley, pp. 157–83.

Massey, D., 1979, 'In what sense a regional problem?' *Regional Studies*, 13, pp. 233–43; repr. in Massey, *Space, place and gender*, Cambridge: Polity, 1994, pp. 50–66.

Massey, D., 1983, 'The shape of things to come', *Marxism Today*, April, pp. 18–27; repr. in Massey, *Space, place and gender*, Cambridge: Polity, 1994, pp. 67–85.

Massey, D., 1991, 'A global sense of place', *Marxism Today*, June, pp. 24–9; repr. in Massey, *Space, place and gender*, Cambridge: Polity, 1994, pp. 146–56.

Massey, D., [1984] 1995, *Spatial divisions of labour: social structures and the geography of production*, 2nd edn, Basingstoke: Macmillan.

Massey, D., 2001, 'Opportunities for a world city: reflections on the draft economic development and regeneration strategy for London', *City*, 5/1, pp. 101–5.

Massey, D., 2005, *For Space*, London: Sage.

Massey, D., and Catalano, A., 1978, *Capital and land: landownership by capital in Great Britain*, London: Edward Arnold.

Massey, D., and Meegan, R., 1978, 'Industrial restructuring versus the cities', *Urban Studies*, 15, pp. 273–88; repr. in Massey, *Space, place and gender*, Cambridge: Polity, 1994, pp. 25–49

May, J., Wills, J., Datta, K., Evans, Y., Herbert, J., and McIlwaine, C., 2006, 'The British state and London's migrant division of labour', Department of Geography, Queen Mary, University of London.

Meek, J., 2006, 'Super rich', *The Guardian*, 17 April, pp. 6–13.

Mensah, K., Mackintosh, M., and Henry, L., 2005, 'The "skills drain" of health professionals from the developing world: a framework for policy formulation', London: Medact: http://www.medact.org/content/Skills%20drain/mensah%20et%20al.%202005.pdf.

Mitchell, K., 1993, 'Multiculturalism, or the united colours of capitalism', *Antipode*, 25/4, pp. 263–94.

Mitchell, K., 2004, *Crossing the neoliberal line: Pacific rim migration and the metropolis*, Philadelphia: Temple University Press.

MORI, 2004, *What is a Londoner? 2nd April 2004*, London: Greater London Authority.

Muir, H., 2006, 'Ken's oil for brooms deal: fuel for us, a clean-up for Caracas', *The Guardian*, 13 September, pp. 1–2.

Nash, C., 2005, 'Equity, diversity and interdependence: cultural policy in Northern Ireland', *Antipode*, 37/2, pp. 272–300.

ODPM (Office of the Deputy Prime Minister), 2003, *Reducing regional disparities in prosperity*, London: ODPM.

OEF (Oxford Economic Forecasting), 2004, *London's linkages with the rest of the UK*, London: Corporation of London.

Olds, K., and Yeung. H., 2004, 'Pathways to global city formation: a view from the developmental city-state of Singapore', *Review of International Political Economy*, 11/3, pp. 489–521.

Oxfam GB, 2000, *Tax havens: releasing the hidden billions for poverty eradication*, Oxford: Oxfam.

Peck, J., 2003, 'Political economies of scale: fast policy, interscalar relations and neoliberal workfare', *Economic Geography*, 79/3, pp. 331–60.

Peck, J., and Tickell, A., 1992, 'Local modes of social regulation? Regulation theory, Thatcherism and uneven development', *Geoforum*, 23, pp. 347–63.

Peck, J., and Tickell, A., 2002, 'Neoliberalizing space', *Antipode*, 34/3, pp. 380–404.

Penniman, H. R., 1981, *Britain at the polls, 1979: a study of the general election*, Washington, DC: American Enterprise Institute for Public Policy Research.

Perkin, H., 1996, *The third revolution: professional elites in the modern world*, London: Routledge.

Perrons, D., 2001, 'Towards a more holistic framework for economic geography', *Antipode*, 33/2, pp. 208–15.

Planning, 2001, 'Fears South may 'hi-jack' science project', 3 October.

Price, C., 1979, 'Recovering the doorstep vote: three-dimensional socialism is now needed', *New Statesman*, 18 May, p. 706.

Pryke, M., 1991, 'An international city going "global": spatial change and office provision in the City of London', *Environment and Planning D: Society and Space*, 9, pp. 197–222.

Pryke, M., 1994, 'Looking back on the space of a boom: (re)developing spatial matrices in the City of London', *Environment and Planning A*, 26/2, pp. 235–64.

Pryke, M., 2005, 'Geomoney: an option on frost, going long on clouds', mimeo, for seminar 'Towards a cultural economy of finance', Open University, 15–16 September.

Pugh, J., 1989, *The Penguin guide to the City*, Harmondsworth: Penguin.

Purdy, D., 2005, 'Human happiness and the stationary state', *Soundings: a journal of politics and culture*, no. 31, autumn, pp. 133–46.

Rawnsley, A., 2001, *Servants of the people*, London: Penguin.

RCN (Royal College of Nursing), 2003, *Here to stay? International nurses in the UK*, London: RCN.

Robins, K., 2001, 'Becoming anybody: thinking against the nation and through the city', *City*, 5/1, pp. 77–90.

Robinson, F., 1999, *Globalizing care: ethics, feminist theory, and international relations*, Boulder, CO: Westview Press.

Robinson, J., 2002, 'Global and world cities: a view from off the map', *International Journal of Urban and Regional Research*, 26/3, pp. 531–54.

Rowell, A., Marriott, J., and Stockman, L., 2005, *The next gulf: London, Washington and oil conflict in Nigeria*, London: Constable & Robinson.

Rustin, M., 2006, '*The long revolution* revisited', paper presented at *Soundings* conference 'Thinking Ahead', London, 3 June, mimeo.

Saghal, G., and Yuval-Davis, N., 2006, 'Cultural difference and integration: terrorism, fundamentalism and multifaithism', paper presented at *Soundings* conference 'Thinking Ahead', London, 3 June, mimeo.

Said, E., 1985, *Orientalism*, London: Penguin.

Samers, M., 2002, 'Immigration and the global city hypothesis: towards an alternative research agenda', *International Journal of Urban and Regional Research*, 26/2, pp. 389–402.

Sampson, A., 2004, *Who runs this place? The anatomy of Britain in the 21st century*, London: John Murray.

Sassen, S., 1991, *The global city: New York, London, Tokyo*, Princeton, NJ: Princeton University Press.

Sassen, S., 1998, *Globalization and its discontents*, New York: New Press.

Sassen, S., 1999, 'Whose city is it? Globalization and the formation of new claims', in J. Holston (ed.), *Cities and citizenship*, Durham, NC: Duke University Press, pp. 177–94.

Sassen, S., 2000, *Cities in the world economy*, Thousand Oaks, CA: Pine Forge Press.

Sassen, S., 2001, *The global city*, 2nd edn, Princeton, NJ: Princeton University Press.

Seager, A., and Balakrishnan, A., 2006, 'Young exiles embrace the Anglo model', *The Guardian*, 8 April, p. 21.

SEERA, 2004, *Perceptions of the South East and its regional assembly*, report prepared by MORI for the South-East of England Regional Assembly, http://www.southeast-ra.gov.uk/publications/surveys/2004/mori_report_july_2004.pdf.

Sharp, H., 2004, 'London: a view from the region of Yorkshire and Humberside', DTI/RSA conference: *London and the rest of the UK: the economic relationship*, 29 November.

Sinclair, I., 1997, *Lights out for the territory*, London: Granta Books.

Smith, N., 2002, 'New urbanism: gentrification as global urban strategy', *Antipode*, 34/3, pp. 427–50.

Smithers, R., 2005, 'London schools still struggling three years after Blair's initiative', *The Guardian*, 19 November, p. 12.

Soper, K., 2006, 'The awfulness of the actual: counter-consumerism in a new age of war', *Radical Philosophy*, no. 135, pp. 2–7.

Sparke, M., 2005, *In the space of theory: postfoundational geographies of the nation-state*, Minneapolis: University of Minnesota Press.

Spivak, G. C., 1985, 'Subaltern studies: deconstructing historiography', *Subaltern studies IV*, ed. R. Guha, Delhi: Oxford University Press, pp. 330–63.

Tabb, W., 1982, *The long default: New York City and the urban fiscal crisis*, New York: Monthly Review Press.

Thrift, N., 1987, 'The fixers: the urban geography of international commercial capital', in M. Castells and J. Henderson (eds), *Global restructuring and territorial development*, London: Sage.

Thrift, N., 1994, 'On the social and cultural determinants of international financial centres: the case of the City of London', in S. Corbridge, N. Thrift and R. Martin (eds), *Money, power and space*, Oxford: Blackwell, pp. 327–55.

Toulouse, C., 1992, 'Thatcherism, class politics and urban development in London', *Critical Sociology*, 18/1, pp. 55–76.

Trades Union Congress, 2002, *Half the world away: making regional development work*, London: TUC.

Treanor, J., 2006, 'Revolution hailed but City warned of a looming fight for supremacy', *The Guardian*, 27 October, p. 33.

UN-Habitat, 2003, *The challenge of slums: global report on human settlements*, London: Earthscan.

UN-Habitat, 2004, *The state of the world's cities 2004/2005*, London: Earthscan.

Vidal, J., 2006, 'Plane speaking', *The Guardian*, 1 November, *Society* section, p. 1.

Walker, D., 2003, 'Mayor urged to back "living wage"' *The Guardian*, 10 March.

Ward, K., and Jonas, A., 2004, 'Competitive city-regionalism as a politics of space: a critical reinterpretation of the new regionalism', *Environment and Planning A*, 36, pp. 2119–39.

Ward, L., 2004, 'Graduate salaries climb to £21,000', *The Guardian*, 14 July, p. 5.

Weir, S., 2006, 'Touching the void', *Red Pepper*, no. 142, pp. 28–9.

Wilkinson, R., 2005, *The impact of inequality: how to make sick societies healthier*, London: Routledge.

Williams, H., 2006a, *Britain's power elites: the rebirth of a ruling class*, London: Constable.

Williams, H., 2006b, 'How the City of London came to power', *Financial Times*, 21 March, p. 15.

Wills, J., 2004, 'Organising the low paid: East London's living wage campaign as a vehicle for change', in G. Healy, E. Heery, P. Taylor and W. Brown (eds), *The future of worker representation*, Basingstoke: Palgrave Macmillan.

Woodward, W., 2002, 'UK graduates join brain drain to south', *The Guardian*, 31 July.

World Bank, 2000, *Cities in transition: World Bank urban and local government strategy*, Washington, DC: World Bank.

Young, I. M., 2003, 'From guilt to solidarity: sweatshops and political responsibility', *Dissent*, spring, pp. 39–44.

Young, I. M., 2004, 'Responsibility and global labor justice', *Journal of Political Philosophy*, 12/4, pp. 365–88.

Zevin, R., 1977, 'New York City crisis: first act in a new age
    of reaction', in R. Alcalay and D. Mermelstein (eds), *The
    fiscal crisis of American cities: essays on the political economy of
    urban America with special reference to New York*, New York:
    Vintage Books, pp. 11–29.
Žižek S., 1997, 'Multiculturalism, or the cultural logic of late
    capitalism', *New Left Review*, 225, pp. 28–51.

INDEX